Cultivating a Movement

CULTIVATING A MOVEMENT:
An Oral History of
Organic Farming & Sustainable Agriculture
on California's Central Coast

Photo: Tana Butler

Edited by
Irene Reti and Sarah Rabkin
From Selected Excerpts of Interviews by
Sarah Rabkin, Irene Reti, and Ellen Farmer

University of California, Santa Cruz
University Library

This book features excerpts from selected interviews from a larger archival collection of complete transcripts, audio clips, and images which is available at:
http://library.ucsc.edu/reg-hist/cultiv/home

To contact the Regional History Project:
ihreti@ucsc.edu or
Regional History Project
McHenry Library, UC Santa Cruz
1156 High Street
Santa Cruz, CA 95064
Phone: 831-459-2847

ISBN: 978-0-9723343-6-5
Printed in the United States of America.

Cover Photo: Juan Catalán holding peppers. Photo: Rachel Glueck.
http://hongkongrachel.wordpress.com

Photo on page 327 by Laura McClanathan:
http://www.asepialens.com/

Photos by Tana Butler:
http://smallfarms.typepad.com/

Cover design by Sandy Bell Design:
http://www.sandybelldesign.com/index.html

Indexing Services by Thérèse Shere:
http://www.shere-indexing.com/

Contents

Harley Farms. Photo: Sarah Rabkin

Foreword

Currently, we in the United States are in the early throes of a revolution—a radical change in the way we think about food. The revolution began a few decades ago, small in scale, on a local level, in a few spots around the U.S. and Europe. It is now spreading into our national conscience. From trendy retailers to reality television, "organic" has become part of our national vocabulary, a household word. Gradually, this "new" kind of agriculture is providing options for produce grown without chemicals as it breaks through the perceptions of elitism and hippiedom and into the aisles of local groceries. For this development in our food industry we can thank, in large part, a group of revolutionaries from the Central Coast of California.

Cultivating a Movement introduces us to some of these revolutionaries. Brought to you by the Regional History Project at the University of California, Santa Cruz's library, this volume shares some of the excellent oral history work that has long been conducted at the university in recording testimony by dynamic local figures. The voices captured here provide a first-hand account from one of the epicenters of the early organic farming and sustainable agriculture movement. In doing so, these interviews are nothing short of essential historical resources in the social, cultural, and environmental history of California. They reflect both the roots of a social justice movement, and a sea change in the way Californians live and work on the land.

This organic revolution is best understood as an "ecological revolution,"[1] a historic shift in the way a society operates within its natural environment. Prompted by significant cultural or economic change, ecological revolutions dramatically alter our notions of an appropriate relationship with nature. In California's past, for example, one might note the introduction of the Judeo-Christian notion of controlling nature that came with European colonization, or the advent of capitalist, market-driven production that came with U.S. western expansion. An ecological revolution can be prompted as well by more specific changes in behavior. The advent of commercial horticulture at the turn of the twentieth century was one such shift. This development

---

[1] Term borrowed from Carolyn Merchant, *Ecological Revolutions: Nature, Gender, and Science in New England* (Chapel Hill: University of North Carolina Press, 1989).

led to changes in crop choices and cultivation techniques, allowing for increased profits and productivity while creating long-term impacts on water, soil quality, and human health. With the initial spread of commercial horticulture came ecologically and socially destabilizing practices, including monocrop cultivation and the need for a sizeable, inexpensive, and mobile labor force. Most notably, agriculturalists introduced chemical pesticides and herbicides into their fields, welcoming them as a safe and scientific advancement in the eradication of pests. Spraying food crops with such chemicals became a culturally accepted practice, changing how we grew food and how we expected that food to look and taste. Now, those in the organic movement are inspiring a new ecological revolution to undo the normalization of pesticide and herbicide use and again change the tide of cultural perspectives on food production.

These most recent revolutionaries—the organic pioneers—have faced a daunting challenge: they have had to break away from that economically successful chemical-based model that, in terms of sheer growth and increased profit, transformed the farming industry. Even after crusaders such as Rachel Carson and César Chávez called public attention to the dangers of chemical farming in the early 1960s, those who chose to eschew a spraying regime were embracing what seemed to be an "old-fashioned" way to grow food, forgoing the protections of modern science. Those outside the movement saw the choice as risky, if not, as Dick Peixoto, one grower in this collection, attested, "flat-out crazy." But what at one time seems "crazy" can become the norm and, as we are seeing now, a national trend.

The stories included in this volume capture the roots of this revolution, and in doing so, help flesh out the most recent chapter in California agricultural history. They add to a story long in the making, beginning when the United States took political control of this region in the mid-nineteenth century and Americans ushered in the Jeffersonian agrarian ideal of a nation of small farmers. The crash of the wheat market by the end of the nineteenth century led to the dominance of horticulture (and subsequently, chemically dependent horticulture) in the twentieth century. As we move through the twenty-first century, the story of California agriculture is becoming characterized by a reconsideration of those earlier cultivation decisions. Those participating in this industry are melding farming with concerns about social justice, human health and global environmental stability. *Cultivating a Movement* brings us the personalities from the front line of this most recent revolution, and these individuals narrate the role of the people behind this dramatic paradigm shift.

My own work in agricultural and environmental history focuses on growers in this same region—the Central Coast—in the early twentieth century, preceding the birth of the organic farming and sustainable agriculture movement.[2] These farmers, in embracing and promoting chemically dependent horticulture, were forging their own revolution. Like the organic farmers that we learn about in the following pages, these earlier agriculturalists developed marketing strategies, wrestled with regulations, and found difficulty in securing a steady and affordable labor supply. They, too, collaborated with university programs and community-based organizations to conquer these challenges. Perhaps most notorious in this vein was the collaborative effort to develop ORTHO pesticides in the Pajaro Valley in the 1910s. But these earlier growers also worked towards their industry's health and longevity. For example, in the 1930s growers in the Corralitos district near Santa Cruz participated in a significant soil conservation demonstration project, working with the US Department of Agriculture's Soil Conservation Service as part of Franklin Roosevelt's New Deal. In this case, growers had to be convinced to unlearn long tried and true practices, such as square planting and clean cultivation in orchards. The techniques were replaced by erosion prevention measures including contour planting and ground cover installation, with the demonstration plots serving to "prove" the effectiveness of such approaches. Some growers resisted, fearing the risks associated with change— would the ground cover compete with water supply, for example? But slowly, these new approaches to stemming erosion took hold, promoted by the USDA and select local growers as the responsible way to care for the land. So in this earlier era, in the same region, we find a burgeoning conservation ethic. This story of soil plots reveals a group of growers beginning to address and rectify missteps previously taken. In recognizing the limitation of accepted agricultural practices, indi-

---

[2] See Linda L. Ivey, "Poetic Industrialism: Environment, Ethnicity and Commercial Horticulture in California's Pajaro Valley, from the Progressive Era through the Great Depression" (Ph.D. Diss. Georgetown University, 2003), and Ivey, "Ethnicity in the Land: Lost Stories in California Agriculture," *Agricultural History* 81, no. 1 (2007): 98-124.

vidual agriculturalists in the 1930s revisited their cultivation choices with the long-term stability of their livelihood in mind.

The most recent generation of growers is also revisiting and rectifying cultivation practices, but their work is imbued with a mission that embraces a larger picture of health and sustainability and greater environmental responsibility. They have used similar demonstration techniques to illustrate that healthy and successful production can happen without the use of chemicals and to inspire local growers to make the shift to organic cultivation. In this book, we bear witness to the ways they promoted changes in the cultivation of our food, and developed a new ethic that asked growers and consumers to live responsibly on the land.

The leaders of this revolution come from diverse pasts—some with their roots in family farming, some in science; some college students, some businesspeople, many simply wayfarers through 1960s California. Their varied paths led them to engage in agriculture but *not* engage in chemical pesticide usage. The conclusions they drew about those chemicals' connections to human health were not initially obvious and then, not easily sold to other growers nor the general public. Facing the resistance they did in rejecting the common practice of chemical spraying, and gambling that their choice would pay off in terms of healthy and marketable crops, the early practitioners of the organic farming business risked much in pursuing this ideal. In doing so, they cultivated a movement alongside their crops.

Cultivating a Movement shares the voices of those who dug their hands into the promise of organic agriculture. We learn about the complexity of this chapter of California's agricultural history by listening carefully to their words. Oral history provides distinctive and invaluable access to the past—first-hand testimony about the history as it unfolded. Within these pages are dozens of documents that help piece together the evolution of this movement, each contributing a piece of the real story as to how the momentum began. The participants in this project share their experiences in choosing this path, each unique but equally pertinent. They share the contributions they have made to this new industry, revealing a complex and integrated movement of growers, organizers, and scientists. Perhaps most valuable to the historical record is the testimony to how they have defined themselves and their movement. With definitions of "organic food" and certification issues still being ironed out in the field, hearing the participants themselves talk through these definitions is historical gold. Each brings a unique and personal touch, but collectively, we can see what narrator Jeff Larkey captured brilliantly: "You can't be what I consider to be an organic farmer and just do it for the money. Because you're not going to be doing it right, for one thing. You have to immerse yourself in it. You have to become part of that web."[3]

Those who wove themselves into that web can tell us much about how major social change is accomplished. Their challenge is far from over, and as a nation we still may be in the incipient stages of this transition in the way we grow our food. However, the voices recorded in this volume remind us how far our society has come since these first pioneers experimented in their "suspect" practices of sustainable cultivation, free from dangerous, carcinogenic chemicals. Here lies testimony to how this revolution got underway.

—Linda L. Ivey, Ph.D.
Department of History
California State University, East Bay

---

[3] Ibid. p. 53.

Acknowledgements

This book came to fruition through the efforts of an extraordinary team of colleagues. I especially want to thank my visionary supervisor, Christine Bunting, Head of Special Collections and Archives at the University of California, Santa Cruz Library, who understood the importance of documenting the history of sustainable agriculture and trusted us to bring this project to completion. Thank you to University Librarian Ginny Steel for valuing oral history as part of library collections. Connie Croker and Jackie Russo steered this project through the choppy seas of the university's financial bureaucracy; their equanimity and skill were crucial to our success.

Project interviewers Sarah Rabkin and Ellen Farmer brought imagination, intelligence, skilled interviewing, and subject expertise to this endeavor. Their involvement extended beyond the interviews to project conceptualization, planning, and outreach. Rebecca Thistlethwaite conducted interviews in Spanish with Florentino Collazo, María Luz Reyes, and María Inés Catalán and brought years of experience with farming, activism, and research.

Mim Eisenberg of Wordcraft and Wendy Ledger of Vo-Type transcribed nearly all of these interviews. They worked to meticulous professional standards. Dedicated student editors Lizzy Gray, Michelle Morton, and Sara Newbold transcribed and/or proofed some of the interviews. We were fortunate to be able to hire Sarah Rabkin to co-edit this paperback collection, and Esther Ehrlich, whose skillful editing in the final phase of the project was invaluable. We are grateful to Tana Butler, Laura McClanathan, Sarah Rabkin, and many other photographers (credited individually) for permission to publish their photos.

Several colleagues and friends helped conceptualize the project or provide contact information. We are grateful to Martha Brown, Steve Gliessman, Robbie Jaffe, Lori Klein, Ann Lindsey, Mark Lipson, Sandy Lydon, Laura McClanathan, Bob Scowcroft, Jerry Thomas, and Carol Shennan.

The Ecological Farming Conference (EcoFarm) provided a major inspiration for this endeavor. We hope this book becomes a resource for the community that participates in EcoFarm each year.

The narrators took time away from their farms and organizations to share their life stories and to review transcripts. We cannot thank them enough. Farmers and activists are some of the busiest folks in the world! We want to acknowledge the farmers, activists, researchers, and educators whom we were unable to interview but whose labors nourish and galvanize this movement. We hope this is only the first of many oral histories of this kind.

—*Irene Reti, Director*
Regional History Project

Introduction

"I tell the world that the organic movement started in California, in Santa Cruz County . . ."
—Congressman Sam Farr, Co-chair of the Congressional Organic Caucus

Monocultured rows of lettuce, Brussels sprouts, artichokes, strawberries, and other chemically grown crops still dominate much of coastal Central California. But tucked away along rivers, bluffs, and canyons, and even within city limits, another agricultural landscape is emerging. The oral histories in this book tell of a synergistic and often visionary web of farmers, activists, educators, and researchers transforming our food system in Central California and beyond. This sampling of narratives is drawn from the first extensive oral history of organic and sustainable farming. It documents a multifaceted and interdependent community of change-makers who speak for themselves, offering a window into the history of a movement.

The Central Coast of California, along with the Pacific Northwest, the Upper Midwest, and the Northeast, has formed one of the hotbeds of the international organic farming and sustainable agriculture movement. The University of California, Santa Cruz's Farm and Garden (now known as the Center for Agroecology and Sustainable Food Systems) constitutes perhaps the single greatest reason this region is on the organic map.[1] In 1967, Alan Chadwick, disciple of the Austrian philosopher and biodynamic farmer Rudolf Steiner, founded the Student Garden Project at UC Santa Cruz on a sunny hillside overlooking the Monterey Bay. Within a few years, a generation of young apprentice farmers had spread Chadwick's French intensive/biodynamic methods and philosophy of gardening and farming across the United States and overseas. Chadwick linked this generation to earlier British and American organic farming pioneers such as F.H. King, who wrote *Farmers of Forty Centuries: Permanent Agriculture in China, Korea, and Japan*, mycologist Sir Albert Howard, who observed traditional farming practices in India and wrote about composting and the health of the soil in *An Agricultural Testament*, and British agronomist Lord Walter Northbourne, who learned biodynamic farming from Steiner and is credited as the inventor of the term *organic farming*.

---

[1] For a history of the Center for Agroecology and Sustainable Food Systems see http://casfs.ucsc.edu/about/history/sustainable-agriculture-at-uc-santa-cruz

The apprenticeship in ecological horticulture at the Center for Agroecology and Sustainable Food Systems continues to be a vital force and has trained over 1200 apprentices in the last forty years. Orin Martin, manager of the Chadwick Garden and longtime teacher in the apprentice program, says in his oral history in this book, "We want them to have the nuts and bolts of how-to; practical and high-order thinking skills regarding organic gardening and farming. And yet, it's more than that. They have an appreciation, respect, a sense of wonderment about the matrix of air, soil, and plants."

But in addition to UC Santa Cruz and its agricultural pioneers, California's Central Coast is home to several nationally prominent non-profit organizations that have shaped the movement over the past four decades. The origins of one of the most influential of the nonprofit organizations lie with Rodale's *Organic Farming and Gardening Magazine.* In March 1971, Rodale sent several staff members from Pennsylvania to California to launch an organic certification program with fifty-six organic California growers. This effort ultimately evolved into the independent California Certified Organic Farmers (CCOF) based in Santa Cruz. By 1979, CCOF had helped push through the first state legislation defining organic standards. In his oral history, (now U.S.) Congressman Sam Farr describes how he worked with CCOF in his role as California State Assemblymember to pass the 1990 California Organic Foods Act (COFA), which established standards for organic food production and sales in California—standards that later became the basis for the National Organic Program now regulated by the United States Department of Agriculture.

The narratives in this collection offer historical understanding of how organic and sustainable agriculture evolved. This is far more than a sector, a trade, or an industry, but a transformative social and political movement, which emerged from the confluence of several earlier movements of the 1960s and 1970s.

It is difficult to overstate the influence of the alternative lifestyle (hippie) movement of the 1960s and early 1970s on the development of organic/sustainable agriculture. Many of the narrators recall living in rural or urban communes. Former UC Santa Cruz Student Garden Project manager Steve Kaffka says of Alan Chadwick's experimental garden at UC Santa Cruz, "It was at the right place at the right time. It was right in the middle of those two centers of the back-to-the-land movement and the alternative culture of the late sixties and early seventies. It was centrally placed and timely."

The brutality of the Vietnam War propelled many young people into antiwar protests and the counterculture. But in addition to marching in the streets, some of our narrators turned to organic farming and gardening as another way to create a better world. Horticulture teacher and ecologist Richard Merrill remembers, "The riots in Isla Vista started. The Vietnam War was consuming me. I couldn't do research anymore [at the University of California, Santa Barbara]. I was protesting all the time, constantly, because the morality of the war was just unfathomable... I said, I've got to do something more relevant than this. I quit. I went and helped start the El Mirasol garden in Santa Barbara."

Not all pioneering organic farmers were urban-raised back-to-the-landers. Narrator Betty Van Dyke grew up with Croatian immigrant parents who refused to farm with chemical fertilizers and pesticides long before the modern organic movement. She, and others like her, provided invaluable expertise and inspiration to the novice "back-to-the-land" organic farmers at the UC Santa Cruz Farm and elsewhere and created a living link back to an era predating the rise of chemical agriculture.

The United Farm Workers (UFW) stands out as another significant influence. The father of former UC Santa Cruz Farm Manager Jim Leap worked as the insurance agent for the UFW. Leap recalls, "I was exposed to a lot of different people and political concepts related to land reform and labor issues. We went to Teatro Campesino plays and we stood in picket lines and boycotted grapes . . . Based on what I'd seen in the picket lines, I thought, wow, if I really want to make a difference, if I really want to be an activist, the best thing I could do would be to find some land, plant some crops, treat people with respect and basically set an example of a viable option."

The sixties and seventies witnessed the birth of modern environmentalism. Biologist Rachel Carson's *Silent Spring*, published in 1962, documented the devastating effects of pesticides on the natural world and sparked legislative action that led to a ban of DDT in the United States by the end of 1972. Bob Scowcroft, who recently retired after a long career, first as director of California Certified Organic Farmers and then as director of the Organic Farming Research Foundation, was introduced to the organic farming movement through his position as national organizer on pesticide issues for Friends of the Earth and anti-pesticide activism remains a strong part of the organic movement.

At the University of California, Berkeley, entomologist Robert Van Den Bosch led the field of biological control, which relies on beneficial insects rather than pesticides to control pests. One of his protégés was entomologist and pest management specialist Sean Swezey, who was hired in 1989 to develop the Farm Extension Project for the UC Santa Cruz Agroecology Program. Swezey began working with other researchers and local, small-scale growers on their farms to analyze the transition of a conventional production system to organic farming practices, including biological control. "Berkeley in the early seventies was a world. Oh, what a world! . . .I still pinch myself that I had that opportunity," Swezey remembers. A few years later, Mark Lipson, now organic and sustainable agriculture policy advisor for the United States Department of Agriculture, entered the environmental studies major at UC Santa Cruz to study planning and public policy and cut his organizing teeth on the battle to stop offshore oil drilling along the California coast.

Lipson also became involved in the co-op movement while at UC Santa Cruz and organized a housing co-op on campus. From there he joined Molino Creek, a farming co-op on the North Coast of Santa Cruz County that specializes in dry-farmed tomatoes. "A lot of people I know in this movement came directly out of the co-op movement," recalls organic farming advisor and activist Amigo Bob Cantisano, who helped organize We the People Natural Foods Co-op in Lake Tahoe, California in 1972. "Our store became the distributing company that is the predecessor, the parent of what is now United Natural Foods... So we're like the parents of that big, huge business. I guess it's the biggest natural foods distributor in the country now. But all that started from these people trying to work together," says Cantisano.

The developing science of ecology also shaped sustainable agriculture and organic farming. Narrators Richard Merrill and Steve Gliessman both were graduate students in ecology at the University of California, Santa Barbara in the late 1960s. Richard Merrill went on to edit the groundbreaking anthology *Radical Agriculture*, published in 1976, and mentored a generation of organic farmers and thinkers through his teaching with the horticulture program at Cabrillo Community College in Santa Cruz County. Steve Gliessman moved to Costa Rica and Mexico to help found the new field of agroecology/*agroecologia*, before coming to UC Santa Cruz to found the UCSC Agroecology Program. Gliessman is part of a cohort of border-crossing agroecologists journeying back and forth between Mexico, Central America, and the Central Coast of California, who have cross-pollinated indigenous agricultural traditions with scientific agricultural research.

These links with Latin America represent an aspect of the historical genesis of this movement quite distinct from the counterculture of the 1960s. José Montenegro grew up in a rural community in Mexico. After studying agronomy in his native country, he emigrated to the United States and founded an organic farming training program at the Rural Development Center in California's Salinas Valley, working primarily with Latino farm workers. This program eventually grew into the Agriculture and Land-Based Training Association (ALBA) Program, which has now graduated farmers such as María Inés Catalán, the first Latina migrant farm worker to own and operate a certified organic farm in California and the first Latina in the country to distribute produce through a community supported agriculture (CSA) program. The barriers to Latino farmers in a still largely English-speaking movement are considerable. "When I went to learn how to develop a CSA, everything was in English," remembers Catalán. "I went to a conference in San Francisco, and we had to sleep in the car. We had coffee and a donut all day because we didn't qualify for the lunches."

The number of women farmers like Catalán who are principal operators of a farm increased by almost 30 percent in the United States between 2002 and 2007.[2] They include women such as Cynthia Sandberg, proprietor of Love Apple Farm. Sandberg's farm is unique in its combination of biodynamic techniques, an exclusive supply relationship with a high-end restaurant, a focus on heirloom tomatoes, a rich offering of on-farm classes, and a successful Internet-based marketing strategy. They include Dee Harley, who, in the coastal village of Pescadero, runs an active dairy and a successful agritourism site. Harley and her staff care for a herd of more than 200 American Alpine goats, crafting the animals' milk into award-winning cheeses.

A new movement is afoot across the United States as young people are taking up organic farming as a career. This changing demographic is critically important, since farmers in the United States are an aging population. Some of these younger farmers complete apprenticeships at the Center for Agroecology and Sustainable Food Systems, study agroecology at UC Santa Cruz, or attend the national Ecological Farming

---

[2] See http://www.nal.usda.gov/afsic/pubs/agriwomen.shtml

Conference, which takes place near Monterey, California each January. "As the nation experiences a groundswell of interest in sustainable life-styles, we see the promising beginnings of an agricultural revival," write the makers of a documentary called "The Greenhorns."[3]

Several of the narrators in this book have received national attention for their contributions to this emerging sector of the sustainable agriculture movement. Juan (J.P.) Perez (age twenty-six), of J&P Organics, was featured in an article on young farmers in the spring 2009 issue of *YES magazine*.[4] Until recently, Paul Glowaski (age thirty) managed the farm for the Homeless Garden Project, a program recognized as one of the Top 10 Urban Farms by *Natural Home magazine*.[5] Nesh Dhillon (age thirty-eight) manages the Santa Cruz Farmers' Markets. *Mother Jones magazine* recently called Tim Galarneau (age twenty-nine) the "Alice Waters of a burgeoning movement of campus foodies." Galarneau represents current cutting-edge efforts, largely youth-driven, to transform institutional relationships in the provision and distribution of food.

This collection also documents the evolution of two commercial crops that dominate the organic landscape of the Central Coast, salad mix and strawberries. The early efforts of Salinas farmers to ship refrigerated boxcars of lettuce to the East Coast are captured in John Steinbeck's novel *East of Eden*. Today 80 percent of the lettuce grown in the United States is from the Salinas Valley. It is not surprising, then, that this area has also become the center of an organic lettuce-growing industry, particularly bagged salad mix. "In important respects, salad mix gave a jump-start to the California organic sector, which then became what is likely the largest in the world in terms of crop value," writes sociologist Julie Guthman.[6] In his interview, Andy Griffin describes his adventures in this burgeoning salad mix market, which began as a special delicacy grown for Alice Waters' Chez Panisse Restaurant in Berkeley, California.

[3] http://www.thegreenhorns.net/

[4] See http://www.yesmagazine.org/issues/food-for-everyone. Also, Mother Nature Network recently ran a feature entitled "40 Farmers Under 40," http://www.mnn.com/food/farms-gardens/stories/40-farmers-under-40

[5] See http://www.naturalhomemagazine.com/People-and-Places/Americas-Top-10-Urban-Farms.aspx. Paul Glowaski now runs Dinner Bell Farm in Grass Valley, California.

[6] Julie Guthman, "Fast Food/Organic Food: Reflexive Tastes and the Making of 'Yuppie Chow." See also Julie Guthman, *Agrarian Dreams: The Paradox of Organic Farming in California* (Berkeley: University of California Press, 2004).

Strawberries began to be cultivated commercially primarily by Japanese American farmers in Watsonville, California in the late nineteenth century. By the 1950s, the same strawberry fields had been intensively cultivated year after year. Fungal diseases such as verticillium wilt posed a significant threat. Berry farmers began to fumigate the soil with methyl bromide and chloropicrin in 1957.[7] Both chemicals are neurotoxins with significant negative health effects.[8] Strawberries are the region's most lucrative crop. In 2007, California produced 87 percent of the nation's strawberries. Fields and fields of strawberries stretch along the coast between Monterey and San Francisco, and while the industry is profitable, there are health and environmental costs to this success.

But if the Central Coast fostered industrial strawberry agriculture it is also the birthplace of modern organic strawberry production. In 1983, Jim Cochran founded Swanton Berry Farm on the North Coast of Santa Cruz County, becoming the first commercial organic strawberry grower in the United States. Researching organic strawberry growing, together with professor of agroecology Steve Gliessman, Cochran says he ". . .went back to good, old-fashioned farming practices, which are crop rotation and adding good soil amendments. Managing soil disease was a problem that Steve Gliessman and I worked out, looking at some of the old literature and coming up with the notion that the Brassica families—plants like broccoli or Brussels sprouts or mustards—might help suppress soil disease."

Despite the groundbreaking advances in organic production pioneered by many of the narrators in the book, today only about 2 percent of the U.S. food supply is grown organically.[9] Not reflected in this

---

[7] See Randall Jarrell, Interviewer and Editor, *Hiroshi Shikuma: Strawberry-Growing in the Pajaro Valley* (University Library, Regional History Project, University of California, Santa Cruz 1986). http://library.ucsc.edu/reg-hist/shikuma.html

[8] Methyl bromide is also an ozone depleting chemical. The United States Environmental Protection Agency has classified it as a Restricted Use Pesticide under the Montreal Protocol on Substances that Deplete the Ozone Layer. It is in the process of being discontinued; however the strawberry industry continues to apply methyl bromide under exemptions granted each year. Now the industry is turning to Methyl Iodide, a carcinogenic chemical.

[9] http://ofrf.org/resources/organicfaqs.html. According to the Organic Trade Association's 2011 survey the organic sector grew from $3.6 billion in 1997 to $29 billion in 2010. The industry grew 8 percent in 2010 while the food industry as a whole only grew 1 percent. The U.S. has 14,500 organic farms covering 4.1 million acres. According to the 2009 crop report, almost 3000 acres of Santa Cruz County's farmland, or 15 percent of the total acreage, is in organic production., up from 800

statistic are the many farms that grow their food sustainably but are not certified organic. These two terms overlap and diverge; the distinctions between them are the subject of heated taxonomic debate. *Organic farming* is a term now regulated through certified organic standards. *Sustainable agriculture* is a broader term and perhaps more difficult to define.[10] We felt it was essential to use both terms to describe this oral history project.

Oral history interviews are informed by the historical moment in which they are recorded. As this project unfolded between 2007 and 2010, so did a remarkable and sometimes alarming series of events. In August 2006, a load of baby spinach from the Paicines Ranch in San Benito County, California contaminated with E. coli 0157:H7 was mixed in with several other batches of spinach being processed. Within a few weeks, over two hundred people fell ill, and many of them ended up in the hospital. Two elderly women and a young child died from acute kidney failure. Organic and conventional growers reeled from the effects of this outbreak, which was ultimately linked to contamination from cow manure.[11]

In March and April of 2007, pet foods that incorporated melamine-contaminated wheat gluten grown in China caused illness or death from kidney disease in several hundred American cats and dogs. This stimulated a pet food recall and concerns about contamination of both the human and pet supply, both in China and beyond. Then came dramatic fluctuations in the price of petroleum and sharp increases in the world

---

acres in 2002. See "Resident visits Washington to Lobby for Organic Industry," Donna Jones, *Santa Cruz Sentinel,* April 10, 2011.

[10] Alternative Farming Systems Information Center librarian Mary Gold explains that, "Supporters of sustainable agriculture come from diverse backgrounds, academic disciplines, and farming practices. Their convictions as to what elements are acceptable or not acceptable in a sustainable farming system sometimes conflict. They also disagree on whether sustainable agriculture needs defining at all." According to Gold, Wes Jackson, evolutionary biologist and founder of the Land Institute in Kansas, was the first to use the term *sustainable agriculture,* in 1978. For an overview of the debates about defining sustainable agriculture see Mary Gold's online publication "Sustainable Agriculture: Definitions and Terms." http://www.nal.usda.gov/afsic/pubs/terms/srb9902.shtml#toc2 The University of California's Sustainable Agriculture Research and Education Program [UC SAREP] argues that "sustainable agriculture integrates three main goals—environmental health, economic profitability, and social and economic equity…Sustainability rests on the principle that we must meet the needs of the present without compromising the ability of future generations to meet their own needs."http://www.sarep.ucdavis.edu/Concept.htm

[11] http://www.aphl.org/aboutaphl/success/12/pages/default.aspx

prices for rice, wheat, maize, and soybeans that caused a series of food riots in the spring and summer of 2008. President Barack Obama was elected and the Great Recession arrived. All of these events shaped both the interviews and the trajectory of the movement being documented.

More information about the social context of these interviews, as well as interview data is provided in the introductions to each electronic transcript available at http://library.ucsc.edu/reg-hist/cultiv/home. The oral histories were conducted according to the best practices of the Oral History Association.[12] We custom-designed topic outlines for each interview and generally asked open-ended questions. Most of the interviews lasted between two and four hours, and were recorded in digital audio. The recordings are archived at the UC Santa Cruz Library. We transcribed the interviews, lightly edited them for punctuation and paragraphing, and returned them to the narrators for their editing and approval.

From the archive of fifty-eight oral histories, we selected for publication here those we felt translated best to the written page. We edited interview excerpts for flow and readability, added some editorial material in brackets, and deleted the interviewers' questions. These excerpted and more heavily edited narratives were also reviewed by the narrators. Choosing which interviews to publish was difficult, as we deeply value the words and achievements of *all* of the narrators who participated in this project. Our intent is to highlight the collection in a manageable, single volume that can be used in classrooms or savored over a cup of tea. Readers are encouraged to peruse the digital archive of 4500 pages of transcript, images, and audio clips.

This book is a tribute to the history of a transformative social movement. We hope it inspires and serves as a resource for students, activists, aspiring and current organic farmers, educators, researchers, farmers' market managers, natural foods retailers and distributors, and the increasing numbers of us who care passionately about how the food we eat is grown and distributed.

---

[12] http://www.oralhistory.org/network/mw/index.php/Evaluation_Guide

Betty Van Dyke

Betty and her son, Peter Van Dyke. Photo: Tana Butler

Born in 1932 to Croatian American farmers in the Santa Clara Valley town of Cupertino,[1] Betty Van Dyke saw her fertile home ground transformed, in a few decades, from seemingly endless orchards to unrelenting urban sprawl. As the matriarch of a family-run fruit-growing business, she has since participated in the region's organic agricultural renaissance, overseeing one of the first California operations to grow and dry fruit organically and playing an active role in the early days of California Certified Organic Farmers (CCOF). And as a member of one of the region's noted surfing families, she built this business while sustaining her love affair with Pacific Ocean waves. Van Dyke Ranch sits at the base of the Gavilan Mountains in Gilroy, Santa Clara County, on a southern exposure well suited for growing flavorful Blenheim apricots and Bing cherries.

[My family] came from the island of Vis in the Adriatic Ocean, off the Dalmatian coast of the former Yugoslavia, what is now Croatia. My grandfather came in the late 1800s, when he was eighteen. He

---

[1] For more on the history of Croatian immigrants in California agriculture see Donna F. Mekis and Kathryn Mekis Miller, *Blossoms Into Gold: The Croatians in the Pajaro Valley* (Capitola, CA: Capitola Book Company, 2009).

[later] brought his family over. They came through Ellis Island, probably around 1919, 1920. The farmers from that particular island all settled in Cupertino. [My grandfather] bought his first land there in about 1922, and started farming with five acres of apricots. The way they progressed in their farming business was the older people that had been there before would buy whatever piece of land opened up next door to them. That's how we got maybe forty-five or fifty acres of land in that area. But not all in one piece. It was all spread out everywhere.

[By 1968] we were surrounded by subdivisions. The biggest farmer sold out first. He owned one hundred acres on Stevens Creek Road. And that went to the Kaiser people, who built the Kaiser homes, the first subdivision there. Then a lot of little farmers sold. And then you had subdivisions going in everywhere, and the property tax was going up. If the subdivision was split into mega-houses, all those houses were paying like three or four hundred dollars, and then they were charging us the same amount. It got to the point to where the farmers were just going broke. And they said, "Wait a minute," and that's when the greenbelt situation came in. We had meetings with the county people, the city people, all the different people that got the property tax at the time. They had to do something for the farmers or they were all going to go, which they all did anyway, eventually.

My father always wanted one piece of land where he could be in one place without hauling the tractor. So [he acquired] 107 acres in Gilroy. It had been prune orchards in the early 1900s and the late 1800s, and very good apricots. And then after that, it became strawberries. Then my father planted forty acres of apricots and twenty acres of Bing cherries. In the '80s, I planted what was left over, into another thirty acres of apricots and a few cherries.

My father had also bought a piece of land that was fifty-four acres, about six miles away, but as he got older, it was too much trouble for him to work it, so he sold it to my cousin, George, and four other partners. My cousin finally bought the other partners out, and then he ran into some problems because he'd overextended himself, so I bought the land back.

I was born in 1932, and grew up on the farm [in Cupertino]. My first job [at about six years old] was picking brush with my grandfather. The second job was picking apricots. You drove through with trucks, no forklifts then. Apricot trees have a tendency to shed their fruit if you don't pick them when they're really ripe. It takes a lot of work to do apricots. We would pick up the apricots that fell on the track, where the truck was going to drive, so that they wouldn't get squashed. We sold our fruit to the canneries in those days. But we dried all the number-two fruit and the ground fruit, which was your best fruit because it was really ripe. So we would put it on trays and wash it with the hose, shake it, make sure it was clean, and then cut it. And then we would sell the dried fruit to the Rosenberg packing house.

As I grew up, I worked more and more in the orchards. You had to sort the cherries before you sent them to the cannery or before they were packed. And then with the apricots, you had to sort them before they went to the cannery, because the cannery wanted perfect fruit. So you took out culls. Otherwise they would dock you. It was funny, because if there was too much fruit that year, they wouldn't want to pay the price. They'd find more reasons to dock you. "Oh, you can't can this because it has a speck," or "You can't can that because it's got a scratch." If you really got mad, you would take your load back and cut it all. [laughs] That's the way it was in those days.

I spent a lot of time with my grandfather in the orchard because he and I were sorters. We had tables that you made that you would carry, a sorting table. You would carry it from one group of trees to another. Your men would pick three rows on each side of the drive that you drove your truck through. First you worked by the hour because you're picking selected fruit. And then when there's more fruit, you work by the box or the bucket. You would get punched for every bucket. That gives you more incentive because you can make more money.

By the time I was twelve, I was doing the books for my father, and the payroll. But I was still working in the orchard with my grandfather. We worked seven days a week, during the harvest. There was no ten-minute breaks in the orchard, or no timeouts. I mean, it's not like there were slave drivers. Everybody worked like that. That was the way everybody in America worked, so it was no big deal.

I [learned by being] there every day. I can't say that anyone took me by the hand. I worked with my grandfather a lot. Everyone you worked for taught you something. I went out in the orchard with my dad. I was on the trucks all the time, and he took me everywhere. I worked for other growers who were relatives, because I was a good worker and a fast worker. I worked for a pear packer. In fact, I used to eat so many pears, the guys in the packing house used to say, "Betty Ann, we're going to weigh you when you go in and weigh you when you go out and charge you by the pound." [laughs]

I worked for the Bogdaniches. I worked for the Marianis. I worked for the Mardesiches. I picked prunes. I packed cherries. I cut apricots. Did strawberries for Mary Ellen, jam. Did work in Libby's Cannery, three or four years with my girlfriends. Good money. Better than working on the farm, because you had labor unions. If you worked on the farm for a dollar, you got a dollar nine at the cannery, and that was good money at the time. Every one of us that was a farmer's daughter worked in the canneries too. All my girlfriends worked in farms. You wouldn't see each other for the summer, because you lived three or four miles away. So we used to write letters.

One year when the prune season was really late in the Santa Clara Valley, they started school two weeks late so that all the kids could pick prunes. Would they do that today? No. They don't even want kids to cut apricots. You have to be sixteen to hold a knife. Silly laws, because one of the best things you learn when you're growing up is to work, and the camaraderie you get when you're working with someone in fruit crops, when you end up throwing food at each other, and playing games, and learning how to work together. You're under pressure, because every crop you're harvesting has got a point where it's going to start dying or getting overripe. It teaches you team sports. You learn to work with people and deal with different personalities. Was there a mentor? There wasn't, really. It was just the whole experience of everybody working together.

In the Croatian culture, men and women, you're equals. If you worked for my father and you were a child under fourteen and you couldn't do what an adult could do (which we all could), he gave you children's wages. But basically he paid everyone the same wage. If a man was getting a dollar an hour, and a woman was working next to him, she got the dollar an hour. There was no prejudice or discrepancy. It was like that with everything we did on the farm, with his ranch, and most of the other farmers too. There were a couple of farmers I didn't like working for because they were really cheap. They would treat me okay because they knew my father, but they wouldn't do as well with their other workers, and I didn't like working for them. Those were the two guys who always got into arguments. Went to court [laughs] with each other. I figured, good company.

The wonderful Okies and the Arkies during the '30s would come every summer to work for us. The Dust Bowl had happened and it was very hard for them to make a living, so they would come to California. It was a treat. I just couldn't wait for everybody to come. They'd come up the driveway with the mattresses on the car, the goats in the back if they had a trailer, or a lot of kids. My father would designate a couple of rows in the orchard for them to set up camp. I'd have a village out there, a little town of people—kids my age, older people, and people that came back year after year. The old folks would come in the summer and work, and the people that were working in the naval shipyards would come on weekends, because they liked being with the family and picking. They were all farmers. It was a lot of fun.

Some of the people had trailers, but most of it was tents. And like I say, mattresses on the roof. At night we'd have music, and people would sing. We had Russians under the tree, [laughs] singing Russian songs. Some of our relatives from San Pedro would come with their fishing boats. Whatever house they stayed at, everyone would go there for dinner. They'd bring all kinds of fish and we'd barbeque. We'd have thirty or forty people there that were all relatives. They'd cook lobster or crab, and then put all the food on the apricot cutting trays, and sit on cannery boxes around the fire. They'd tell stories and sing into the night, out-of-doors. It was an amazing way to grow up.

[In college] I started out with music and business, because I was a pianist. Then I went into occupational therapy, because I liked science and math. But I did a lot of anthropology, because I liked anthropology. Then I started surfing. I spent the last couple years in college living over here in Santa Cruz and surfing. But I still worked on the farm, in the summer.

I had seen a picture in *Sunset Magazine* when I was young—I loved the water and the ocean—of somebody standing on a surfboard. I always wanted to do that, but I didn't know where, how, and didn't know they did it here. My friend Al [Wiemers] had known the Van Dyke brothers, and a couple of other people over here in Santa Cruz that surfed, and brought me over one day, and I went tandem and caught a wave. And that was it. [laughs] That was '53. But we still worked on the farm every summer, except the summer when I graduated college. Then I went to La Jolla for the summer to surf. I told my father, "I'm going to do this, this time for myself," and he said, "Okay."

My two oldest sons were both surfers. A lot of the guys who are farmers here are surfers. They are at the farmers' market but they all go surfing in the morning. [laughs] When you're a farmer, time's your own, really. Unless it's harvest, and then you've got to be there no matter what. So it's a good combination. And that's why a lot of these kids that are farmers are surfers too.

I never thought that [farming] was what I was going to do, because my father was doing it. I didn't think about becoming a farmer until he got older. He really needed someone to take over, and I knew what to do. Before he died, he looked out the window and he saw the dryer yard and he said, "Well, I see you know what to do now. It's all up to you." So that was it.

It wasn't like I was new to the farm or anything like that. I got myself twelve dozen gopher traps and started trapping gophers right away. [laughs] He hadn't been able to do it for quite a while, so we had quite a gopher problem. It was no big transition in my life. It was just something I'd done all my life. It's ingrained. It's not, how did you learn? It's just there. You don't even think about it.

My father grew up in Croatia and he was in charge of feeding the family while my grandfather was gone. He learned how to grow everything organically, because, of course, there were no pesticides there. Whatever pesticides they had, they were natural. He was always getting the prize for the best garden. He was proud of that. So when he came to America he wasn't into chemicals. Farmers were using a pesticide that you would spray on the apricot trees. It would stunt the growth of the tree but make the fruit get big. They discontinued its use after a few years because it was carcinogenic. He used to go out and warn the farmers next to him, "Don't get that on my land." And he always said to me, "Don't spray anything in the orchard that you can't go in afterwards and pick the fruit off the tree and eat it." With the cherries, basically all he ever did was an oil base. He never sprayed with any chemicals.

Alan Chadwick was having his classes [at UC Santa Cruz].[2] Word got around that we didn't spray our cherries and all of a sudden these kids were showing up, going, "Can we pick after the crop is over? We heard you don't spray." I said yes. I remember one group. They were on some American Indian trip. When they got through picking in the orchard, they were in front of my father's window where he always

---

[2] Van Dyke is referring to the early years of the UC Santa Cruz Farm and Garden, when Alan Chadwick taught apprentices his blend of biodynamic and French Intensive gardening in the garden near Merrill College at UC Santa Cruz.

used to like to sit and drink a glass of wine. They put their hands in a circle [laughs], and were chanting something. My father just loved it.

And that's how we got into the organic movement. All of a sudden, stores just started asking us. Community Foods wanted to buy our cherries. So I started with ten rows of apricots that we sprayed only with the same sprays that my father used, which was blue stone and lime, which is a copper-based spray. We did those ten rows, and we had twenty-five feet on each side that you could spray with Rovral, I think, a fungicide that we used at the time. That's how we started. And Ken [Kimes] and Sandra [Ward] were my first inspectors. We've been friends ever since.

We were certified the third year that the CCOF was in. A few years after that I took the whole place out of production because we had been hit with a heavy rain. My son had left the fava bean crop up too high. I went skiing and I told him to plow it under. He didn't plow it, and the fruit brown-rotted because of the moisture, so I had to take it out of organic that year. But in those days you would get certified within a year if you hadn't used any chemicals. For those of us that [are] all of a sudden hit with a really bad year where it rains a lot and it destroys your crop, I wish they [the organic certification agencies] would give us leeway to just spray a fungicide. Fungicides don't hurt anything. They never touch the fruit. They do it before the blossom is on the tree. It doesn't affect the ground, the worms. It dissipates. And that's the only thing we have problems with. Three years ago I had put another section into organic. It's been treated organically except for this fungicide. I waited the whole three years, and then last year I went out to the orchard and I could see the rot starting to happen. I told Peter, "Okay, go spray with the fungicide. It's a whole twenty acres. I can't let it go to waste." And I thought, God, another three years to go through this [certification] again. It's just pretty hard to do.

I've belonged to CAFF [Community Alliance with Family Farmers] since it started, and I've belonged to CCOF [California Certified Organic Farmers] since the third year. The Ecological Farming Conference—the third year they had it I think we spoke on the "Successful Farmers and What They Do" panel. It used to be up in La Honda, and then they moved it over to [Asilomar]. We went to it every single year until the year my son got married. We used to go for several days. We'd always drive instead of [staying overnight at the conference] because we were always doing something at our place. It was pruning season. I don't go regularly like I did before. And it's changed. Because, before, all of us led the groups. You'd go to the conference and there would be one of us there teaching you how to trap gophers, or teaching you how to dry fruit. Amigo Bob [Cantisano] was teaching you how to make compost. Everybody was teaching each other and they were all learning from each other. I mean, everybody came. You'd see the same people every year. Now it's more business. It's more marketing. It's big. You still see the same people, but now there's a lot *more* people, you know?

A Blenheim apricot, when it's ready to go, you get a little breeze—there was times when I'd be sleeping in the trailer at the farm, and the wind would start blowing in the morning, and I could hear: plop, plop, plop. You got to get up early in the morning before the pickers get out there, get the fruit that's fallen. At one time, all the Santa Clara Valley grew the Blenheim apricot, because it was and still is the best dried fruit. That's what my parents always had, and so that's what we grew. And it is the best apricot.

We never shipped it at that time, because it was very fragile. When I took the farm over from my dad and I started doing organic, I did start shipping them. I have shipped them all over the United States. The receiving end has to be patient and understanding. You have to pick them with a certain firmness and greenness. Once they have color they will ripen. They will have flavor. But you can't ship them when they're full color, because they'll ripen in a day or two, and they're too soft. We had pretty good luck with Raley's, and Nob Hill, and Whole Foods, and Texas Health. All the places back East: Roots and Fruits, Rainbow. I've turned it over to my sons, so I've forgotten all the people that we sell to. But most of them now are all part of Whole Foods anyway, which has gotten huge.

What's destroyed the Blenheim apricot business is the Turkish imports. Your Blenheim apricot was your best dried fruit, and we used to dry fifty to ninety tons. But now it's down to a handful of us that dry maybe less than a hundred tons. The Turkish apricots are so cheap. They come in at fifty cents a pound. In fact, the packers are not giving us the money that we should be getting for the dried fruit, and that's what's hurting our business a lot. I'm trying to sell more fresh [Blenheims], which is good, because once [people] get them, that's what they want. Most of the other [fresh apricot] fruit that's out there is not good. They're bred for shipping. And they don't make good dried fruit either. But the Blenheim, it's the queen of the apricots.

You could always tell whose trees were blooming [in the Santa Clara Valley of my childhood]. It was a field of snow. First of all, you would have the almonds. Those would be the first trees to bloom. When I would look out the upstairs window, the almonds would be first. Then the second thing to bloom would be the apricots. Then after the apricots would be the prunes. Those two were really close together. Then the last thing to bloom would be the cherries. And the thing that I always found very interesting was the almonds bloomed first and were harvested last. The cherries bloomed last and they were harvested first. The cherries are little tiny things, and the almonds have to get hard.

Cupertino and Los Altos is where the ground started slowly sloping up. We were just on a high enough level in a two-storey house to be able to look over everything. There were only a couple of peach orchards. Peaches bloom pink. There were just a few. I couldn't see to Alviso, so I could never see the pear trees bloom. Most of those were by the bay, because pears like a lot of water. They had big fields there. Row crops were over there, too.

You just had San Jose, Santa Clara. Cupertino was a grocery store and a post office and a bank. (Well, that came after the war.) Sunnyvale was a little town, and then Mountain View. Los Altos was really tiny, like a block. Then it was Palo Alto, Menlo Park, Redwood City and those places. But they were all tiny. They weren't connected with subdivisions like they are now. So you had a lot of room between, a lot of orchards between the little towns, so there was really a differentiation.

There was a railroad track that ran along San Francisco to San Jose. We always took it. When we were working in the cannery, we'd take the train to San Francisco to go shopping. We'd take our cannery checks, [snaps fingers and laughs] four or five hundred dollars in the pocket, and go to San Francisco and buy school clothes. We'd take the train. But along the railroad tracks was also a road that you could drive on, that went through the back of all the towns, like Sunnyvale. You could go straight on that road and go through Mountain View and then Palo Alto. And Mayfair was a little town. It's part of Palo Alto now, but that was a little tiny town.

Sometimes Easter vacation would come really early, and the trees would bloom late. Paradise. Absolutely. You would drive up Highway 17 from Santa Cruz to go back to the valley, and before you reached the summit you could smell the blossoms. Then you reached the summit. There was no smog, so you could see all the blooms. It was really beautiful. Blossom Hill Road—they named it that because it was a little elevated and it went along the edge of the valley there, along the foothills. You turn right when you get to Los Gatos and go on Blossom Hill Road, and you had this beautiful view of all of the valley to the left. In fact, some of the major pictures that were taken of the Santa Clara Valley at that time were from Blossom Hill Road.

In the spring in Saratoga they would have the Blossom Festival. People would come and you would sit on a hillside, and there was a little amphitheater, a little tiny, tiny Greek amphitheater. And from where you were sitting on the hillside you would see the valley in bloom in front of you, and this little Greek amphitheater where they had ballet. I saw Jascha Heifetz play the violin there when he was about seventeen or eighteen, because he lived there. We did that every year.

I think that the whole Santa Clara Valley should have been preserved. It's one of the finest places to grow fruit in the world. When they put every freeway in, they never used a stick of dynamite. You know, they go down thirty feet to make an underpass. It was all scoop.

I didn't know you had to work in a garden until I got to Santa Cruz. I didn't know about hardpan. You put a seed in the ground in Cupertino, it grew. The whole valley was like that. The whole valley. It was the richest valley in the world.

Amigo Bob Cantisano

Photo: Saxon Holt

Amigo Bob Cantisano is perhaps best known as the founding organizer of the annual Ecological Farming Conference, which is the largest and oldest sustainable agriculture gathering in the Western United States. One of the most widely experienced and influential figures in California organic agriculture, he was a cofounder of the California Certified Organic Farmers (CCOF) and collaborated in the production of an early organic products trade journal, as well as a natural foods store and organic foods distribution company. When Cantisano farmed organically near Yuba City, California, his search for a reliable source of organic soil amendments led him to found Peaceful Valley Farm & Garden Supply. He later established the first organic agriculture advising business in the country. Operating for more than two decades now as Organic Ag Advisors, he has consulted with hundreds of small and large growers, advising both organic farmers and those making the transition from conventional farming.

My ancestors were the first white people in California. My direct ancestor was the lieutenant in the de Anza Spanish march that came across from Monterrey, Mexico, and settled all the places along the coast of California—San Diego, Santa Barbara, Los Angeles. They went to Monterey, where de Anza set it up as the capital of [California], and then sent my ancestor, Joaquin Moraga, out to establish the next settlement, which turned out to be San Francisco, or Yerba Buena. He was the commandant of the Presidio. That was in 1775. So I'm either a seventh-generation or a ninth-generation Californian, depending on how you count.

I grew up in a family that ate pesto. And until I was probably in my twenties I never really saw that out in society. People didn't eat fava beans. Artichokes were kind of rare. I remember all that stuff was part of our regular heritage. And big diverse salads. I mean, everybody was eating head lettuce in the fifties and sixties. My grandma made salads with radicchio. I just remember there were these very colorful and real diverse kinds of salads. Thank goodness she was such a chef and gardener.

My great aunt and uncle, when I was a kid, owned a farm in Lodi. They grew wine grapes and cherries. They were in the wine grape business during Prohibition. They sold all their grapes for sacramental uses, because that's the only way you could legally sell grapes during that period. Some other parts of my family were in Sacramento involved in running illegal wine. It was part of their deal. That was the Cantisano part of the family. And that's how, apparently, the Cantisanos and the Moragas[1] met up.

I used to go out in the summers and hang out with my great uncle Anthony. I remember he would let me ride or drive the tractor. I got to go out and do the harvest with them in cherry season, but I was really slow as a kid, so they would let me do things like moving the buckets around, or moving the ladders, or working in the packing shed. And then as I got older, we would drive the truck or tractor around, picking up the boxes.

In 1968 and 1969 I lived in communes, mostly in the City, in the Haight.[2] At one point I lived in Pacifica in a commune. That's when I started doing gardening, because it was like, starvation time, and the Diggers[3] were giving away free food, but the only thing else that was

---

[1] See http://www.wjasper.com/Moragahistory.html for history on the Moraga family.

[2] The Haight-Ashbury is a district of San Francisco named for the intersection of Haight and Ashbury Streets near Golden Gate Park. The district was a center of the hippie movement, which swarmed "The Haight" during the Summer of Love in 1967.

[3] The Diggers were a radical community-action group of actors in San Francisco. They took their name from the Diggers in England in the 17th century, who promulgated a vision of society free from private property and all forms of buying

really around was USDA surplus food, which was a bunch of crap. We would take [the food from the Diggers] gratefully. That was before food stamps. We didn't have any money, so we started gardening, tore up the backyard and started growing stuff, partly because I had a little experience from when I was a kid, and partly because I was just determined to do it. And then there was one other kid who had grown up and lived on a farm. He clearly knew something about growing stuff. So the two of us ended up being kind of the commune's gardeners.

In 1968 I started working, first as a dishwasher and later as a prep chef, at the Good Karma Café in the Mission, which was this natural foods vegetarian café—among the first, I am told. Fred Rohé owned this café. Fred started the first natural foods store, a supermarket. It was called New Age Natural Foods. It was a converted Purity building, which was kind of like the predecessor of Safeway or one of those chain stores. It was in the Haight, on Stanyan Street.

Fred Rohé is, I think, the founder of the whole natural foods movement. He was a businessman, a brilliant guy. At that point in time, the only place you could buy natural foods pretty much was the red carpet health-food store pill-shop phenomenon. Nobody had yet gone into the natural foods business; in fact, they didn't even use the term "natural foods" yet. It was "health foods." But Fred opened this mini-supermarket of natural foods. I worked in that store for a few months, part of the time in the produce department, and met some of the organic farmers.

And then I worked at his Good Karma Café—right across the street from the mission that my ancestors helped settle. That's where I first started eating natural foods. Brown rice, I'm sure I'd never eaten before that. Or stir fries or tofu. So in that circle of people, plus the hippies I was hanging out with, there was all this interest in "natural foods." That was the up-and-coming thing. And at some point *Organic Gardening* ended up in my hands and I read it. I thought that was pretty cool.

Fred Rohé is also the fellow that started the first natural foods distribution company, which was called Pure and Simple. That was based in San Jose. They were the first people to put Tom's natural tooth products in California. They were the place where you went to buy Dr. Bronner's [soap], because they would stock that kind of stuff. Later the other natural foods distributors started, like

---

and selling. During the 1960s, the San Francisco Diggers opened stores that provided free food, medical care, transport and temporary housing. They also organized free music concerts and works of political art.

Veritable [Vegetable], S.F. Common Operating Warehouse, Rock Island Foods, Fowler Brothers, etc.—but Pure and Simple was the first.

And then in the summer of 1969, I moved to a commune in Mendocino County, where again the same problem: no money. We started growing stuff, and at some point in time the word *organic* kind of crept into the conversation. I realized my grandmother had been an organic gardener, but no one ever talked about that. But she made compost and got chicken manure from a neighboring chicken ranch, and grew fava beans, and as far as I knew didn't use any pesticides. All the weeds were controlled by hand. She was a de facto old-world gardener. But then it became sort of like: "Oh, this is what organic gardening is. Oh, it's: don't use chemicals and use these natural fertilizers and such."

And then going to [the first] Earth Day [celebration in San Francisco] in 1970 solidified it in me: Oh, there are these risks to the pesticides and the way people have been growing stuff in monocultures. At that point, I didn't really understand all that, although the DDT issue and the information that Rachel Carson had written stimulated me very much.[4]

What was also springing up around that time was the buying clubs and the food co-ops and the collectives. A lot of people I know in this movement came directly out of the co-op movement. "Food for people not for profit" was a big motto. I actually took that quite seriously.

So in May of 1970 some friends of mine were moving to Lake Tahoe, and they invited me to come up. We went to the health food store. And it was really disturbing because it was like going back four years. It was just this plastic pill shop. There were no natural food stores. So a friend of mine, Eddie Kitchen, and I said, "Let's put up a notice on this bulletin board and see if people want to be in a buying club or a food conspiracy."[5] We held this meeting, and fifteen people showed up in somebody's living room. Everybody said, "Yeah, let's do this." So Eddie and I said, "Okay, we'll take a truck to the city and buy some food for the group." Because I'd worked at New Age Natural Foods, I'd been to a bunch of these distributing companies picking up stuff, or they'd driven stuff in, so we had some idea of where these things were coming from.

---

[4] *Silent Spring* was published by biologist Rachel Carson in 1962. The book is widely credited with helping launch the environmental movement. *Silent Spring* facilitated the ban of the pesticide DDT in the United States in 1972.

[5] "Food conspiracy" is a term that described various organizing efforts in the early 1970s among neighbors who pooled resources to purchase food directly from

Organic Food Club accumul.
grows into distributor

We took a run in a pickup truck down to the Bay, and came back home with a pick-up load of food, and distributed it out amongst us. And that just grew like wildfire. So I basically became the food club coordinator. Well, that started us going to visit farmers. At that time there was only one farmers' market in Northern California, the Alemany Farmers' Market in San Francisco. So we would go to that market, and there were a few people there who would say they were organic. So we preferentially would buy from them, although we would buy from conventional people too, because there was almost no organic stuff around yet.

Our little buying club grew to the point where we were renting an eighteen-foot refrigerated truck and hauling it twice a month. We were doing the food distribution on Sundays in a friend's plumbing warehouse. We'd break all the food down and everybody would show up. We had hundreds of members at some point. So then we were like, "Well, gee. Maybe we should drive around and go visit farms." So then we started looking for farms between Tahoe and the Bay Area, Sacramento County and Yolo County.

We were the first people to buy rice from the Lundbergs[6], like the first! The only product they had was hundred-pound sacks of short-grain rice. So we would drive up to Chico. And then, Knudsen's and Heinke's were the two juice companies. And almonds. So we would go get these different Central Valley [products], and then stop off at Howard Beeman's farm near Woodland, which was doing sweet corn and melons, and a few other small farms. And this trucking group, Santa Cruz Trucking, would come over from Santa Cruz and they'd bring some produce up. This is the informal start of what became all of these businesses that then started trucking and coordinating stuff.

Then the buying club started moving into more like a business of transporting and selling organic foods wholesale. And our food co-op moved into a storefront and became a retail business. That was on the North Shore of Lake Tahoe, in Tahoe Vista. It was called We the People Natural Foods Co-op. It was because we founded the store on the fourth of July in 1972. I was a co-manager with a group of other people, which were basically all the members of my commune. It went crazy from there. We were driving a truck once a week, and then twice a week to

---

farmers and small distributors.

[6] See http://www.lundberg.com/farming/philosophy.aspx for more information on the philosophy and history of Lundberg Family Farms.

the Bay. And we helped get a natural foods bakery going in our town. They needed all this fresh flour from Giusto's. So we started a distributing company that was an offshoot of the retail store. And then that thing just got a life of its own, because there was nobody yet doing that kind of stuff. So we would hook up with this guy, Gilbert, who used to drive down from Nancy's Yogurt in Oregon in this '49 Chevy panel truck with the yogurt stuffed in the back under dry ice. He could barely make it back and forth. And we'd meet him on I-5. Somebody would say, "Oh, do you know about this chicken ranch over here in Sonoma?" and we'd end up over there buying eggs. Bit by bit, all of those businesses got solidified, but at the beginning they were all fly-by-night, seat-of-the-pants.

Our store became this distributing company [that] is the predecessor, the parent of what is now United Natural Foods. We the People merged with Sierra People's Produce, which was a friend of ours, Lee May. He was doing produce exclusively, and it wasn't really working for him to haul produce. He had a trucking route but he didn't have a truck. He rented a truck. And we had a propane-powered truck. So the co-op and he merged and became Sierra People's Warehouse. And then eventually Lee took over that business completely. Then he merged with a guy, Michael Funk, who had Sacramento People's Produce. They ran that for a while, and then Michael moved to Nevada City and he named it Mountain People's Produce. And that was that business for a long time until it merged and became part of United Natural Foods.

So we're like the parents of that big, huge business. I guess it's the biggest natural food distributor in the country now. But all that started from these people trying to work together.

The more I visited farms, the more I realized I really wanted to do farming. Running food around was cool and it was exciting. We made a little bit of money doing it. I could pay my bills. But we were gardening. We were living in a commune in Truckee at this point, and we had about a two-acre garden. We were selling stuff to the co-op. But that's a very limited climate.

One of my roommates, Gus Rouse, was a beekeeper. He'd grown up as a farmer in Yuba City and he wanted to go back to farm, so he and his brother Monty started doing commercial beekeeping. Their grandpa owned seven acres of land that wasn't being used, that he had bought near Yuba City. He hadn't moved onto it yet. I wanted to learn about beekeeping, and the intrigue of farming vegetables, nuts, and fruits for the co-op was very big. Gus and Monty knew how to farm and they needed help with the bees, so we made a pact to work together.

It turned out Grandpa Rouse was the wise old organic farmer, although he had given up the organic farming thing when we met him. When I met him in the seventies, he was already almost eighty years old and he'd farmed way before chemicals. He'd farmed his whole life in the Santa Clara Valley or up in Los Gatos. So he knew all about how to do this stuff: "Grandpa, how do you grow a cover crop?" "Well, here's how you do it." "How do you make compost on a farm scale?" "Well, here's how you do it." "How about rotate crops?" He'd kind of laugh at us and go, "Oh, you guys are crazy," but he knew that we had this market with the co-op. He wanted to see his grandsons stay in ag, because everybody was leaving.

So we had the bee business, which eventually got huge. We called it Starr Farms. It had six hundred hives by the end of the first year. We did pollination, honey, bee pollen, royal jelly and produced queens and packages of bees for other beekeepers. Then we had this seven acres or almost eight acres of vegetables, plus about two acres of mixed fruit and nuts. And we leased a fifteen-acre almond orchard, with a few walnuts.

Grandpa Rouse was our mentor. We were farming but were pretty lost, at least I was. He would come up to visit his grandsons and his new property. He was still farming in the Santa Clara Valley, across from IBM there. He would just kind of shake his head. He would go, "Well, at least my grandsons are farming. They got this hippie kid farming with them." And then sometimes we'd have ten or fifteen friends come and help. It's a little chaotic. You know, this is the early seventies. We're definitely in the hippie, party mode. So people are running around mostly naked on his farm. He had a little bit of a hard time with this. [laughs]

We made some mistakes. Oh, my God, *horrendous mistakes*—like the time we— Of course when you're broke, anything that looks free or cheap looks awfully attractive, right? So what was free in our area? Rice hulls and chicken manure. All you had to do was pay for the trucking. Well, that just was like, "Cool, man. We've got instant fertilizer." Well, little did we know that— I mean, even though Grandma taught me about composting, I didn't know about the downsides of using raw manure yet. And I knew *nothing* about using rice hulls, other than, God, they were just this wonderful fluffy stuff. So we got this just humungous amount, like a hundred yards or something, and we spread it on one acre. It ended up almost a foot thick. You could barely plow it in.

Nothing would grow. I mean, just nothing! We'd put a plant in there and it turned yellow and just sat there. It was just like, oh, my God! The weeds started growing eventually, because there were a lot of weeds in the rice hulls, as it turned out—maybe some in the manure, probably in the rice hulls. And then Grandpa just walked up one day and went, "Well, you can't do that."

We were so naïve, and we were literally alone in our attempt to farm organically. There weren't any organic growers. In our county, there was one other fellow who was growing organic blackberries. He had an acre and a half of blackberries he'd inherited from his grandparents and decided to grow organically. But it was nothing like growing melons and tomatoes. There were no mentors. Grandpa would show up a couple of times a month in between farming, and come up to visit, and just kind of walk around. He was really nice about it. He never chewed us out, but he was very straightforward in saying, "This isn't going to work, and here's why."

But we didn't let that stop us. We kept going forward. That acre or so was pretty much a waste for a couple of years. It eventually broke down and it was farmable. It turned into some nice soil but it took a couple of years.

I had a great experience the next year. We decided to lease this walnut orchard. Again, we were young and we were thinking we could do anything. We had the vegetable farm, and we had a chance to lease a walnut orchard and an almond orchard. The almonds were pretty straightforward, so we went ahead and leased them. It was really cheap. Almonds were worth *nothing* at that point in time. These were not very healthy trees, and somebody had pretty much walked away from fifteen acres of them.

And then there was this really nice walnut orchard down the street. The lady's husband had died three or four years before, and it was basically abandoned for three or four years. We didn't have a law yet for organic. You could call anything you wanted organic. It was a beautiful walnut orchard and it had flood irrigation, a lot of really good things going for it. So we went and leased this orchard. And Grandpa said, "Well,"—and he was a walnut farmer—"I don't think you're going to be able to grow these without pesticides." There were two bugs that they had, the walnut aphid and the walnut husk fly. So I went, "Well, I'm going to go down and find out the latest from the farm advisor." I'd never been to the farm advisor's office yet, [the] University of California farm advisor in Sutter County.

I'll never forget it. The guy's name was Larry Fitch. Now, I'd already farmed organically. I'd already been marketing and doing organic food for, I don't know, six years or something by that point. I walked in there and I told him what we wanted to do. And I remember coming home and telling my girlfriend, who became my wife, "Kalita, there's no way. We're going to have to spray these. There's no way to farm these without chemicals. There's just no way to do it." Larry had brainwashed me in the space of probably forty-five minutes. He had convinced me that it was impossible to farm these things without chemicals, so we were going to have complete crop loss. Well, by this point we'd already signed the lease. So it was kind of like, okay, we'll go down this road and see where it goes.

And honest to God, a miracle happened. That's all I can call it. We're out there working the place one day, and this guy drives up in a pickup and he says, "Are you Bob Cantisano?" I said, "Yes." He says, "Well, I hear you're trying to farm these without chemicals." I said, "Yes." He said, "Well, I'm Bill Barnett from the University of California." I said, "Yes, Bill. That's what I'm trying to do." He said, "Well, you know, I'm working on the IPM [Integrated Pest Management] pear pest management manual, and we're trying to take pictures of beneficial insects, but we can't find any in the pear orchards. We can't find a lot of these species. We know they're there, but it's been sprayed so much that we can't find these beneficials. So we're wondering if we could look around your place and see if we could find some beneficials to take pictures of?"

Well, it turned out my farm ends up with about fifteen pictures in the pear pest management manual (they don't describe that it's on vegetables or walnuts) including the cover picture of this lacewing. Anyway, so he's there. He brings this photographer and he's standing out there for a couple of days and gets all these beneficial pictures. Well, meanwhile, I didn't know hardly anything about beneficial insects at that time but Bill knew tons. So I started picking his brain. "Hey, Bill. What are we seeing here? What is this?" And he'd say, " Well, that's a Nabis. And that's a syrphid fly and this is what they eat."

Well, I started talking to him. I said, "You know, Bill, they've got me convinced here that using chemicals is the only way I'm going to be able to farm these walnuts." He said, "Oh, do you want to do tests and do some trials?" I said, "Yes. Great. What do you want to do?" He said, "Well, I've got this idea about trapping that we use out in the forest for this beetle that attacks pines, a different species but it's a similar idea. Maybe we could try that out here." I said, "Whatever you say, Bill. Let's give it a road test."

Well, it turned out it worked. It wasn't just that. We did a bunch of different things. But we did figure it out. We grew beautiful walnuts with very low cullage rate. Even the first year we had very little damage. And Bill was like, "Yes, this is cool." He wrote a paper about it and got it published, because at that time there was nobody that was not spraying walnuts. Chemicals were the gold standard and we proved that it was possible to farm walnuts without chemicals. It was not possible to just let nature takes its course and get a good crop. There was the need to be proactive. So we did the mass trapping, released lacewings to attack both the husk fly and the aphid, cultivated around the trees to disturb the overwintering walnut husk fly pupae that are near the soil surface under the tree, released a species of wasp that attacked the walnut, sprayed every tenth tree with a molasses-based attractant bait combined with an organic insecticide and other IPM techniques.

Well, then we became the university test plot. They would try to do these no-spray projects and then the farmers would get nervous in the middle of the season and go spray them out and they'd lose their trials. Finally they'd found somebody who wasn't going to go spray them out. They could actually do the research (because I guess they didn't find any other organic farmers at the time). They found somebody who would say, "I won't spray it. I'll take the damage, the risk."

So we ended up becoming the test plot for this little parasite that they brought over from Iran, this little wasp for the aphid. The aphid was a big problem, a big, big problem. Everybody was spraying for it. We had problems with it too. We'd found out that certain predators would feed on it, but there was nothing that was specific on the aphid. It was an introduced pest from Iran, because they brought walnuts over here originally from Iran, or that area, with the aphid pest and they didn't come over with their biocontrol parasite. UC professor and biocontrol entomologist Robert van den Bosch was working on this parasite from Iran. He turned out to be my guru, thank God. Bill introduced me to van den Bosch, and then it was like, oh, my God, now we have someone who is champion of what we're doing, knowledgeable, and enthusiastic.

So Van came back with this parasite and said, "Okay, you guys. We're going to put it in here," and it turned out it totally wiped out the aphid in our orchard in only one season. UC researchers spread it around other places around California in abandoned walnut groves, and it completely eliminated walnut aphid in California in the space of four years. Bang, biocontrol equals not pesticides. That was in the mid-seventies. At that point farmers were spending over a million bucks a year on pesticides for that aphid alone, in walnuts. And that one introduction, which cost like five grand, it was like nothing. They sent Van over there. He found these parasites. He raised them up. They put them out. The things multiplied like crazy and it eliminated pesticides for that particular pest, which has never been a pest since.

Okay. So then all of a sudden now we're getting written up in journals, and I know nothing about this stuff! (laughs) I'm an "expert." I'm not an expert at all. I'm just a farmer, but it gave me a lot of encouragement about the possibilities, because I realized that if you found the right person with some science knowledge, lots of things were possible. But at the time, even then, there were so few people doing organics that people didn't really care about organic farmers. The university had not yet given any credence of any significance, and there were just a few individuals like Van and Bill Barnett. Bill wasn't fully an organic guy, but he believed in that IPM idea, and so he believed in beneficial insects. But then we'd try to talk to them about soil management and they'd glaze over. They had no clue about soil and plant health. They're entomologists.

So partly how I ended up becoming an "expert" was because I couldn't find anybody who could give me all the pieces of the puzzle. They would give you little segments. They knew about the weeds; they knew about the pests; they knew about the predator; they knew about the disease; they knew something about soils—but nothing, really.

You couldn't buy cover crop seed in the seventies. You could not buy it except at Northrup King, an international seed company that sold primarily crop seeds such as corn. They had two winter grains, oats and barley, and purple vetch. That was the only place in California you could actually buy cover crop seed. Grandpa Rouse was telling me about all these varieties he used to grow in the twenties and thirties: "You should grow Indian sour clover. It kills the gophers." I'm like, "Okay. Let's go find *Melilotus indica*." We researched all over the place. And of course there was no Internet, so it's phone calls, mostly. And they're saying, "No, I don't have that."

So we go down to the university library in Davis to learn about cover crops. Man, was that disappointing: there hadn't been a publication written on cover crops since the fifties. Not a single thing written at the university since, like, '52 or something. And then we found this treasure trove, this book from 1928 that had all the stuff. Oh, it was an amazing book. Oh, it's *the* book! In fact, it's still the best book on cover cropping out there. Nothing's been written that tops it. It was first published in 1927. It's called *Green Manuring: Principles and Practice* and it's by Adrian Pieters. It was sitting in the reference section of the UC Davis library and you couldn't take it out, so I went and photocopied every page, which in 1975 cost a fortune. But fortunately, in 2004 a company in India reprinted it. So it's now in print. And it's still the best book. I have a friend, Edwin McLeod, who wrote a book on cover cropping in the seventies titled *Feed The Soil,* and he took a bunch of information directly out of the Pieters book, because there were no current resources at the time.

So here we had Grandpa Rouse who had farmed during this era, right? 1927. He knew all about cover crops, and this book. But no cover crop seed. It was kind of like, "Oh, great. This is a great idea, but where do you get this stuff?" So then we talked the Lundbergs into growing us some vetch and peas. That's what started that whole deal. One of the first things that [the organic farm-supply business I started], Peaceful Valley Farm Supply, did was get into cover crop seeds. That was its niche. Because nobody else was doing it.

Farming
↓
distribution
↓
Farming
↓
Entrepenure

Andy Griffin

Photo: Tana Butler

Andy Griffin's farming roots reach back to California's 1970s organic farming renaissance. His practical education took place in a series of jobs on farms—including an organic garden that supplied produce to Alice Waters' Chez Panisse, the Straus family dairy, and Warren Weber's Star Route Farms in Marin County. After stints as a produce distributor, he eventually established Riverside Farm with partner Greg Beccio. The proceeds from that large salad greens business funded the creation of Happy Boy Farms, now run by Beccio, and eventually helped Griffin establish Mariquita ("Ladybug") Farm, which he runs today on twenty-five acres in Watsonville and Hollister. Griffin and his wife, Julia Wiley, sell much of their produce through a community supported agriculture venture called Two Small Farms.

My formal agricultural education pretty much ended with high school. I was in the FFA program in Carmel High, the Future Farmers of America—raised beef steers, went to the fair, took ag classes through FFA. I went to UC Davis [University of California, Davis], originally with the idea of studying range management. When I got there, I discovered, to my dismay, that you actually had to pass science classes. Beyond that, range management then consisted, as far as I could tell, in learning all the sprays you would need to get rid of the sagebrush off the Great Basin Desert and replace it with alfalfa. It didn't really interest me at all, so I left.

I figured I'd learn how to farm by working on farms. I worked on the Straus Dairy. They're pretty well known now for being one of the first organic dairies, but they weren't organic then. They were in the throes of really hard times. The dairy business was in deep shit. And that was a really interesting time to be there. Really nice people. Ellen Straus—she's passed away now, but she's the woman who started the Marin Agricultural Land Trust, and that grew out of the concerns that they had just trying to keep their farm going in West Marin. I lived with her son, Albert, who runs Straus Dairy now.

It's hard work, working on a dairy. It's really brutal. So I went back to the university, but I got a degree in philosophy. I already had in mind that I would have to learn how to farm by working on farms. I wasn't going to go back to UC and study their version of farming. That didn't appeal to me on any level. I didn't see it leading to anything. But I wanted to go back to school and, you know, meet girls and whatnot. And I was so out of touch. I just figured I would study whatever I felt like. It never occurred to me that you don't meet the bubbliest girls in a philosophy class.

I got a lot out of it, and at the end of the bachelor's program, I was anxious to get back [to outdoor work]. During the summers, I worked on farms and on the weekends I worked on farms. What turned me on to organic farming was a summer job. I worked for the Cargill Corporation one summer in Davis. That was just straight-ahead fieldwork, a vast, conventional farming operation. We were [growing] sunflowers. I like being outside. I like farm work. I wasn't very happy with the whole Cargill program. They would even spray the sunflowers with herbicides. They all died at the same time. I mean, it's real serious conformance. That didn't appeal to me. But it was a job, and I got paid, and it was outdoors, and it was in the Sac[ramento] Valley.

The next summer, around 1979 or 1980, when I was looking for a summer job, I got an offer off some student employment center for a job on an organic garden. It paid six bucks an hour, and I thought, why not? What turned me on there were actually the people. This farm was run by a fellow named John Hudspeth, who was working as a chef at Chez Panisse in Berkeley. He came from a lot of money. They had a large piece of property there near the airport out on Garden Highway. And he wanted to have a biodynamic, French intensive garden and sell the stuff to Chez Panisse.

One of the things he was doing at Chez Panisse was sourcing stuff, and he saw a way to combine it all together. He hired a woman named Karen Montrose, who came up from the Golden Door, which was a health spa down in the hills behind San Diego. She came up to put this garden together for him. John was not about to go out there and pull roots or something like that. That was not his style. But he really had the vision of what he wanted to do. And so what I found was this very different farming system than I'd ever encountered before.

By the time I was working on this garden, [Chez Panisse celebrity chef] Jeremiah Tower was already gone. But they had a real established sense of who they were already. I was never really part of the Chez Panisse community. That was as close as I ever came, working on that farm. I got pulled into the orbit, heard a lot of stuff, was exposed to a lot of people. And the thing that I liked about it was, number one, having fun was a big part of what they were all about. And, two, they were so turned on by what they were doing. That enthusiasm is infectious.

My own experience working with vegetables was very limited. I didn't know anything about vegetables. And I was exposed to many different kinds of crops which I had never heard of before, which most people back then had never hear of, like arugula, which is almost a commodity now. Nobody knew what that stuff was. There must have been 150 different things in that garden. We grew everything from valerian root to borage flowers, the edible flowers. All this stuff which became a rage later on, they were doing.

My background up to that point had been in cattle-raising, which is a very unsentimental business. It didn't take me very long working there to look around and see that they were essentially trying to re-create for themselves sort of a Provençal lifestyle. I hadn't gone to work there because I wanted to have a Provençal lifestyle. I mean, I didn't know from Provence.

The first year I worked there, I heard that they lost twenty thousand dollars. The second year that I worked there, I heard that they lost twenty-six thousand. The numbers aren't exact in my head. It could have been thirty thousand they lost the first year and forty thousand the second year. And as somebody who did a lot of the work there, I looked around and thought, you know, this double digging, this massive planting of all these vastly different kinds of crops for sort of putative [Rudolf] Steinerite principles, this is all having a really hip, faux-Provençal lifestyle, but this is not making any money. If you operated a specialty organic vegetable farm on production principles [instead], you might make money. You might not have as much fun, but you might make money.

So when it was time to leave school at Davis, I left that farm. I knew that I was going to be leaving Davis, and I had gone back down to the

Straus Dairy in Marshall and spoken with Ellen Straus. She knew a lot of people, and she and I had always gotten along well. I had been maintaining a garden at the house when I was living on the dairy there. And she said, "You really enjoy vegetable gardening. You should think about being an organic vegetable farmer. You're not going to make much of a dairyman." Because I didn't want to milk cows. I was concerned if I got a job milking and I was used to the money, that's all I'd ever do. I didn't really want to follow in my grandfather's footsteps and milk cows for my entire life. I had declined to become a milker [at Straus Dairy], and I earned less to do the other chores just because I didn't want to slip into that. She knew I didn't have any future in dairying.

I went back and I said, "Well, Ellen, can you think of any place where I can go get a job in an organic vegetable farm in West Marin?" because I loved living in West Marin. It was just wonderful. It was like a park. And she said, "Well, you should go down and talk to Warren Weber in Bolinas." So I went down there and got a job. Went straight from Davis to Warren's farm, Star Route Farm[s], and spent the next five years there.

That was a fascinating experience for me, exactly the kind of thing that I needed, because Warren was really running the farm on business principles. It turns out Warren had a lot of money, too. When I was working at Warren's, he was married to Marion Rockefeller, who would be Laurance Rockefeller's daughter. They have more money than God. Of course, Warren would not have met Marion if he didn't have a lot of money to begin with. I mean, you don't meet girls like Marion Rockefeller at QuikStop.

[But] Warren ran his farm as a business, not a hobby. From my perspective now, having run a farm for a long time and supported myself off it, the luxury I can see that Warren had, the only luxury he allowed himself was that if he made bad decisions and his farm went bust, his family didn't suffer. In a way, that really freed him up to make business decisions from a fairly fearless point of view. But his farm had to pay for itself. They were not going to the grant or the foundation or the trust fund. And that was really a curative experience to working for the vegetable farm [with Chez Panisse], because you make different decisions when you make them from an economic standpoint. Sometimes you're not making decisions about the very best way things can possibly be done, but if you're basing your decision-making process on what *can* be done, you're going to be around. You're going to be sustainable. I loved it there, and I stayed there for five years.

One thing that I learned from Warren that has really been useful later on was his conception of the farm as employing people year round. You could only achieve some degree of stability as a business if you have a stable crew. You get a stable crew by not firing them at the end of every season and starting over again. At the end of the first season, when most other organic farms at that time were small enough [that] they would have just said, "Well, it was great. See ya next summer," what Warren said was, "We don't have enough work here for the whole winter. I certainly can't pay you guys to sit around. But what I'll do is I will pay you for four days of work a week, whether or not we do any work."

And it turned out that the farm flooded that winter, so I got the best end of that deal. There were five of us working there at that time, and we were there in the spring. On the very first day we could get into the ground, we were there. He wasn't out looking for a crew. That was a very important thing to me, and I've tried to run my farm on that sort of principle ever since I had the opportunity to make my own decisions.

I learned anything that I do know about farming from him, and also a vast amount about marketing. Because at that time, in the early eighties, organic farming was still really hippy-trippy, and there was a cultural bias against salespeople. So many of the problems that organic farming had were cultural problems. Warren used to laugh because he went into town one day, into Bolinas, to deliver food to the Bolinas People's Store, and some customer at BPS accosted him and demanded to know if he was farming for people or for profit. It's kind of hard, now that Whole Foods dominates the scene and whatnot, to remember that there was a time when you had to put up with that kind of idiocy. You can't just give everything away.

Warren's father was the kind of ad executive whose accounts included Anheuser Busch and Colgate Palmolive. Those were his accounts. I think Warren's father was responsible for "The Pause That Refreshes" or something. He was a heavy hitter in the advertising field. And one of the most fun things that we would ever do is we would be out there hoeing weeds on Star Route Farm[s], and this was when Warren worked in the fields with the other four of us, and we would design ad campaigns for what it would take to bring organic beyond this little niche, hippy market out into the big wide world.

Some of these ad campaigns were sort of cynical riffs on the culture that we were living within. So we had the "Pure as the Driven Snow" campaign, where we would envision the different video shots and the hyperbolic promises about how—you know, the degree of purity. If somebody else was eliminating any kind of chemical contamination, we would have to outdo by eliminating any sort of spiritual contamination. This would mean that we would promise the customer that we were only harvesting these crops with actual Buddhist monks. Since some of our friends were at the Green Gulch Farm right down the road, this stuff was all based on something.

I really enjoyed that. It was a kick in the pants, and it also was probably the equivalent of a bachelor's degree in marketing, because we learned a lot about marketing from listening to Warren talk. One of Warren's brothers was the kind of consultant who was working for the Republic of Congo, bringing in a hundred thousand Chinese laborers or something to build a vast dam, these mega-international kinds of projects. Talking to Warren was different than talking to people who had no conception. You got a much more 3-D vision of the world around you. One of his very close friends is Orville Schell, who is the dean of [the] Graduate School of Journalism at Cal [UC Berkeley] now, and who was a Chinese translator and scholar. Another one of Warren's close personal friends is Bill Niman of Niman Ranch. Niman and Schell were our neighbors and his friends. And so it was an intellectually charged milieu.

One of the things that I did while I was on the farm was set up a delivery route to restaurants in San Francisco. So I was spending part of the time driving in San Francisco. I had an epiphany one day. Warren had a great big International Bobtail. It's a truck that's small enough that you don't have to have a special license, but it's got eighteen gears, so it's a real pain in the ass to drive in city traffic, a very big flatbed. I would take this in and do these deliveries.

I do deliveries in San Francisco now. I look back on that and I just shake with fear. I would never, ever send anybody into San Francisco in one of those trucks. It's totally unsafe to be doing deliveries in these great big trucks—downtown, narrow alleyways and whatnot. On steep hills, you're trying to change the gears. You don't see anybody driving these things in the city anymore. It's all these Iveco cab-overs, Isuzu cab-overs, these little, short delivery trucks.

But we were doing a delivery to the Campton Hotel. The delivery route called for me to go up Washington Street and make a left onto Stockton, I guess, go through the Stockton tunnel and then deliver to Campton Hotel down there by Union Square. So I was going up the street. I was in Chinatown. It was midday, and that particular day, [along] the entire length of that street, the right-hand lane had been ripped up by a city crew. They were putting some sort of gas main in or whatever, and so all the traffic was over on the left-hand lane.

It was slow going up the hill. There was a little alleyway right before Stockton, and some Chinese kid driving the same kind of truck as I had, had been trying to back his truck out of the alley, and he had dropped the rear duals down an open manhole. He had effectively shut the street off, because the nose of his truck was in a dead alley, the duals are down a manhole, there's a ditch on one side—I mean, you're not going anywhere—and they couldn't figure out how to get a tow truck in there. It was a total nightmare.

I had to back down street after street back through Chinatown. And all these people were crowding around, all those little old Chinese ladies in a rush to go someplace and not paying any attention at all. I had to keep on letting out the clutch and putting on the brake, and I was so freaked out and nervous about it, my right leg was starting to seize up, there was so much tension there. A cop finally came along and saw the problem. He was swatting these people out of the way so I could get through. When I got down to Columbus, I pulled over. I flipped the blinkers on. I went into a bar. I sat down and had a couple of beers. My leg was actually so stiff with stress at that point, I couldn't move it, hardly. This took forty-five minutes to an hour, to go three or four blocks down this hill.

I thought, I'm working on farms because I want to live out in the country, and I'm in Chinatown in the middle of the day, driving an overweight Bobtail backwards down a hill? Something's terribly wrong. I said—you know, readjustment. I went back and I told Warren, "Find another driver. I'll keep working as a driver until you find another one. But find another one, because I need to do something different." So he found another driver, who lasted a couple of days. The guy said, "That's nuts." Warren went out and bought a little Isuzu cab-over, and it all worked out from there.

I started out working with Chez Panisse and harvesting these little organic greens, these baby vegetables that were a fad for a while—but especially things like arugula, the little lettuces. When I was working with them, these things were completely unknown. I went to work for Warren Weber in Bolinas, and a woman was working there named Wendy Krupnick. She started there just a few weeks before me. Wendy had been a waitress at the Bay Wolf Restaurant over in Oakland, where they were doing a menu that was seasonal, California-based.

So Wendy had a lot of the same sort of experiences I did, but from the other side of the equation. She and I had talked to Warren about, "Hey, you should grow these baby lettuces, man. This stuff kicks ass. They want them out there." And Warren was still, "No, we are growing food for people. It's got to be for profit, but we are growing food for people." And we used to laugh about that because Warren came from a really wealthy background. He didn't want to see organics become an elitist sort of crop. He wanted food to be grown organically for everybody. So we were selling food for people. We were selling big truckloads full of potatoes and cabbages and cauliflower, all the staples.

And it became apparent to me at that time, having sort of been first exposed to organic through John Hudspeth, that hippies were just dreadfully conservative. Even by the early eighties, I mean, hippies were already getting old. They only wanted what they'd always had. They didn't want to try anything different. They just wanted it organic. Kind of boring and tedious. John had his issues, but he was a lot of fun, and he was really into cooking. He'd come trotting out in the garden early in the morning. He was going to be making breakfast, so he'd be dressed in full chef drag, with a little white hat and whatnot. He would go out, and he would pick the peas, and then he would rush back in, and he would do this fantastic little thing with a little bit of handmade butter and these peas. He'd bring some out for us. I thought, God, this food is fantastic! And then you go to Bolinas, and it's all hippies, and it's got to be brown rice and lemon juice and a cabbage and a great big chunk of tofu. It just wasn't the same.

When Chez Panisse lost so much money on little projects like John's, they started looking around at getting more professional growers to step in and fill this niche. Their restaurant was rocking out, and they were especially selling a lot of these salads. So they sent Sibella Kraus over. She was working at the restaurant at that time as a forager/procurer, whatever. She is now the empress of a nonprofit called Sage, and she started the farmers' market at the Ferry Plaza up in San Francisco.

Anyway, she came over [to Star Route Farms] one day, and we had a field of lettuce—red-leaf, green-leaf, butter, romaine—twelve-inch centers, forty-inch beds in front of the office there. The plants were about four and a half, five inches tall. She stepped out of the car and introduced herself. She was from Chez Panisse. We were there talking, and she looked at the field, and she said, "Well, I want to buy the lettuce." Warren explained to her very patiently that it would be weeks before that lettuce was mature, and she said, "No, I want it that size." Wendy and I knew all about it. We'd been doing this for a while already. But Warren wasn't really too hip to it. He was so hardcore he said, "Well, you know, when these are full size and mature, I'm going to be able to sell these for fifty cents a head. I can't sell them to you for less just because they're small. I mean, I've already put all this work into it." And she said, "Oh, I'll pay fifty cents a head."

So we cut these things. They were tiny. We loaded them into recycled Styrofoam grape lugs we picked up. We were putting seventy-two heads per box, which works out to thirty-six dollars a box. Warren was dumbstruck. We put more value into the back of a little Ford Courier than we had in this great big International Bobtail. He just said, "Lord! We gotta switch gears here."

Even by that time, the organic market was [changing]. There was starting to become more production online. It was still so limited. There were more farmers, but there weren't more wholesale accounts, and there weren't an appreciably larger number of customers yet, and so there was an over-supply relative to the market. So, it was getting tougher. He saw this as an opportunity to open up into a new market, and one that made sense for the farm because we were close by the Bay Area, and he had two people on staff who felt very comfortable and confident about this.

So we charged into that, and we started taking stuff over to Chez Panisse. Chez Panisse had a really instrumental role in gluing all of this together, because we wouldn't have known what to look for or where to find the seed. Chez Panisse helped people bring the seed over. They helped people get all of this started. So they were a seminal influence in many ways. We started doing quite a bit of business with Chez Panisse. It became a considerable account, although by no means the farm's only account.

The great thing about selling to any restaurant is that after you've sold to the restaurant for a little while, there is inevitably a dispute between the chef and the sous-chef, or the chef and the owner. The chef leaves, and the sous-chef gets a battlefield promotion. And if you've been good to everybody, the chef who leaves takes you with them, and the sous-chef who comes up through the ranks keeps you on. So one account turns into two, turns into four. If you're being good to everybody down to the dishwasher, there's a mathematical progression built into your relationship, and it just spreads.

At that time, the need on the part of the restaurants was way out ahead of the supply. It's now a cliché. They interview a chef, and the chef inevitably today says that their cuisine is based around local, seasonal, organic. I mean, this is a complete cliché. These days, chefs have the opportunity to really do that. Back then, they had the desire, but they didn't have the connections. There wasn't the infrastructure at all. There were farms doing organic produce, but they were not in the same culture as the restaurants. They were coming from a completely different perspective.

This would have been '82, '83, '84, '85. The restaurants didn't know where the farms were. The farmers didn't know where the restaurants were. It was even more profound than that, because a lot of the farmers thought that the restaurateurs were a bunch of effete little French fags, and a lot of the restaurateurs thought that a lot of the farmers were a bunch of bumpkins, and so, there were all these levels of interference. If you just showed up at the door, and you're well spoken, and you've driven to them and you say, "I'm an organic farmer, and I want to deliver directly to you," it's like, "Lord! Where have *you* been all my life? Here, have an espresso. Would you like a shot of grappa in that?" It was great, and I got every account that I asked for.

Warren saw how popular these salads were in the restaurants. You still couldn't throw the stuff away in the organic world, because the organic world was still ruled by arch hippies, who were mostly concerned about not being poisoned by capitalism. There wasn't a role for this stuff yet. But we reasoned that so many people in the East Bay had been eating these salads at Chez Panisse and at Santa Fe Bar and Grill that maybe there would be a retail opportunity for this stuff opening up. You couldn't see the stuff in the farmers' markets yet. And there weren't the same number of farmers' markets as there are now. We were working with Monterey Market by this time. Sibella had helped us find some other accounts. We were working with Bill Fujimoto in Monterey Market in Berkeley, and we approached him with the idea of trying to sell the baby mesclun salad greens through Monterey Market. And he said, "Sure."

So we started planning on doing a retail side to this. We thought it would be really big. We thought, well, in order to make it really big, what we ought to do is we ought to get Alice Waters in on the act, because she has got the publicity mojo to kick this into high gear. So we called her up, set up a meeting, we drove over, and the whole way over there, Warren and I are talking about how we're going to set this thing up. It didn't occur to us that she might not want to do this. Because we saw this money that's so obviously there for the taking. How could you not want to take it? I mean, there it was, this whole product. How often are you gifted with a product that's never been done before that you know goddamn well will sell? Not very often.

We drove over there, and we had a little meeting at lunch with her. She listened to us, and she said, "You know, you're absolutely right. This could be huge." She says, "I don't really want to be a part of it. No offense. My restaurant has been more successful and busy than I ever could have imagined. I have a very young daughter. I'm in over my head. This could be huge, but I kind of like things small. I can give you the names of some people who might be able to help you with packaging and things like that. Go for it!"

She was real clear. I remember thinking to myself, Man, the woman's nuts! Of course, now I realize that actually, she's kind of screwy sometimes, but she's deeply sane. I look back on that, and she was sane enough to know how much money could be made and just not try for it, just figure, hey, things really *are* better on a small scale.

One of the names that she gave us was a friend of hers, Todd Koons. Todd came out, talked with us, and he was hired by Warren to put together the packaging. Todd crafted this whole clamshell approach that we were going to use, because, if you go and cut a lettuce, it starts to wilt right away. There was no technology at all for washing this stuff, spin drying it. It was difficult to create the technology until you had the market. But it was difficult to build the market without the technology. It was a slow process, and the clamshell approach was the first way we went.

We harvested whole heads, reasoning that they would wilt slower if we cut whole heads and tried to make a whole blended salad. So we packed these little clamshells with four kinds of lettuce, with tufts of mizuna, chervil, arugula, things like that, put in there with a mesclun label, stuck it out there. I remember Todd looking at this the day that we finished, saying, "This is gonna be huge. But not like this."

We took it to Monterey Market. They didn't sell very well. Bill Fujimoto took us aside and said, "You know, the problem here is people don't perceive the value yet. If you raise the price, these will sell more." So we raised the price, and they started selling more. Bill is a profound thinker. There isn't much about produce that he doesn't know, or about marketing that he hasn't learned in the trenches. He was aware of the fact that at that time that product still had sort of an elitist spin to it. It was something you ate when you went out to a fancy restaurant. If you price it low enough for just people to eat, that was degrading the image of the product at that time. Nowadays it's sort of funny to think like that, because nowadays you can get the stuff at McDonald's. It's really changed. But back then, that was an intelligent marketing decision.

While I was down in Southern California, Todd Koons started TKO Farms. And when I was down in Southern California, Earthbound Farms, which has got a big name now—they had their little tiny farm in Carmel Valley—they started buying raw product from Star Route Farm[s] and other places, and washing it to make their salads. Dale Coke was the first person I'm aware of who figured out how to spin dry it well enough.[1] The standard at that time became a three-pound food-service pack of mixed baby lettuces. So there was a lot of change going on in a short period of time.

---

[1] See the oral history with Dale Coke at: http://library.ucsc.edu/reg-hist/cultiv/coke

Jeff Larkey

Photo: Carlie Arnold

Jeff Larkey spent part of his childhood in the Carmel Valley of California, and in Davis, California, working summers in the fields and processing plants of the conventional agriculture world. In the late 1970s, Larkey moved onto a commune on Ivy Lane in Live Oak, an unincorporated, then still somewhat rural area of Santa Cruz County, where he and his fellow commune members grew basil and garlic, as well as other crops, on four acres of land and transported them via bicycle to sell at a neighborhood farmers' market. These crop choices helped establish a taste among local community members for fresh pesto. In 1981, Larkey left that commune to farm along the fertile floodplain of the San Lorenzo River in Santa Cruz. His farming operation was certified in 1985 by California Certified Organic Farmers (CCOF), and eventually became Route One Farms, which now leases sixty-five acres of land in several locations in Santa Cruz County.

I [came] to the Santa Cruz area to go to Cabrillo College to get involved in the solar technology program there, which was run by Richard Merrill,[1] who also was developing the horticulture program. I didn't automatically get into farming. I wasn't really sure what I was going to do. But I had moved onto this little commune in the middle of Live Oak. We were growing all our own food and we had milking Jersey cows, and goats, and chickens. It was four acres in the middle

---

[1] See the interview excerpt with Richard Merrill in this book.

of Live Oak. Right about that time was when the California Department of Food and Agriculture was developing this direct marketing program and so a bunch of local farmers tapped into that. And we started the first farmers' market. It happened to be two blocks away from where I lived, at the Live Oak School. We were bike-carting everything over there because we were so close. That was 1978, 1979.

In 1979, we'd take twenty bunches of basil to a farmers' market and bring ten of them home, because nobody knew what to do with basil in 1979. Just in a matter of a few years, we were bringing fifteen cases of basil to the farmers' market and selling them all in two hours. It was just one of those things where you saw something explode right before your eyes. That was a farmers'-market phenomenon. It wasn't something that happened in the stores. It was something that happened in farmers' markets. People were exposed to something that they hadn't been exposed to before because of that small-scale agricultural situation. People were experimenting with things. On a small scale, people could experiment with what they're eating. They just happened to like pesto once they tried it. Not only did they like it, they liked it a lot! Then they told other people about it. It was an unbelievable explosion.

And garlic was big. The first commercial crop I ever grew was garlic. That was the very first crop I ever sold at a farmers' market. We made garlic braids. We grew the dried flowers and we braided them into the braids. We put a big rack of garlic braids up. We made what was for us a lot of money. It doesn't sound like a lot of money now. But back then we were paying sixty dollars to rent a room in this house. Now it's a little different. But in the seventies it was a different time. You didn't have to make as much money as you do now. Now you got a mortgage to pay or whatever. You got to be pulling in thirty-thousand-dollars-plus-a-year minimum just to break even.

[Now] a lot of what I grow goes to Whole Foods, more than half of what I grow. Because now I'm eighty-five acres, I can't just supply the local farmers' markets and local food stores. I've got a lot more going. Every day we're picking eight to twelve pallets of stuff. So that gets shipped. We do Whole Foods. We do Veritable Vegetable, a distributor in the Bay Area. And then Coke Farms, who brokers my produce—they're growers as well and have a cooler that everything gets shipped out of. They're in San Juan Bautista. It's in San Benito County. It's right next to Highway 101, so it is really easy for the big refrigerator semis to pull in and load up a few pallets and take it wherever it needs to go. Whole Foods is a huge player in the flowering of organic farming and organic produce. They're a nationwide, becoming almost an international outfit. They do their share as far as educating the public about regional produce and allowing people to put their money where their mouth is in terms of supporting things like that. They potentially have the predatory instinct that any large corporation can have. Especially being a public company, they're beholden to shareholders.

I don't own any of the land I farm. I've always leased. The longest piece I've leased, on Ocean Street Extension [in Santa Cruz], I've been leasing on a year-to-year basis for twenty-six years, no longer than one year at a time as a lease. It's always under threat of getting sold or developed because of the value. It's a primo location, and it's right next to town. It's alluvial silt there. It's right on the San Lorenzo River, the ancient flood plain. It's twelve feet deep of silt, the most ideal soil you can imagine. To me, that's the value. It is an irreplaceable location because of the soil and the climate. There really are very few places like it on the planet.

To be a farmer is to be a lot of different things. You have to be a good manager. You have to deal with people. You almost have to be driven, because it's not always easy. I'd say one out of every six years is a losing year, for one reason or another. I'm not a tree-fruit farmer or a perennial farmer of any kind. I do mostly row crops. Do some fruit trees and perennials. But for the most part, what I'm doing is successive, basically weekly plantings. Everything I grow pretty much gets planted every week, so I have a continual supply through the season. Your customers demand consistency. If you run out of something, they're going to go to someone else, and they may not get you back. That was a really hard lesson to learn early on. You plant something and think they're going to keep taking it. But if you miss a couple of weeks, they start getting it from someone else. And if that person stays consistent, they just keep getting it from them. So consistency is a really key thing.

[You must have] good people skills, because if you're employing people, you want to be respected, and you want to be respectful of your workers. They're pretty much what keeps your business going. You have to speak Spanish. Most of my workers are professional farm workers. They have to be, because I grow so many different things and everything takes a different pack. You have to know how to do that. You can't just come and work on this farm for like a month or two at a time. It's not that kind of operation. It's got to be someone who really wants to do that kind of work and is willing to work ten hours a day, six days a week. That's just the way it is.

I've got close to thirty employees. They work eight to nine months of the year, the majority of them. A lot of them go back to Mexico in the winter. They've got their own little operations going back there, farms or cattle or whatever. That's why they know what it's about. They don't have to be told how to work. We're usually harvesting until one o'clock, or sometimes later. We try to get that done as early as possible in the day. And then the rest of the day they are moving irrigation pipe, or weeding. Weed control is the biggest thing for us. It's a lot of labor, because we don't use herbicides and weeds are always a problem. We grow a lot of things that are growing really close together. We can't even get them with a hoe sometimes. You have to go in with your fingers and get things out. That's usually what the last few hours of the day is spent doing. [Weeds are] competition for the nutrients and air and space and water. Ideally you'd like a weed-free situation. It's nice to have them around to cover up the ground if you're not planting it, at least temporarily. But you don't want to let them go to seed, because the old saying is, "One year's seed is seven years weed." An amaranth plant can produce half a million seeds. One plant. So you're stuck for a few years if you let that thing go fully to seed.

Being open to new things is pretty darn important, because no matter what you do, no matter how good you think you are at it, eventually it's going to change. The nature of nature is change. What works this year isn't necessarily going to always work. So you have to be open to that. You've got to be up on changes in the natural environment, and educate yourself about the options that you have, because as an organic farmer, you're limited in what type of options you're going to do. If you have a soil disease, if you're an organic farmer it's really difficult to get rid of once you get that disease. A lot of what you do is preventative. So trying to maintain a very healthy soil, which in turn creates a healthy plant—it's a complicated thing to do. I can't go out and fumigate my soil if I get a bad soil-borne disease, a bacteria or a fungus or something like that, it's really difficult to deal with. So you are dealing with things on a microbial level. I mean, that's basically what organic farming is. You want to create beneficial microbes, is what you're doing. In a way, you're kind of a microbiologist. You may not be super educated about those things, but in a certain way you have to pay attention to them.

They increased the production so rapidly after World War II. It wasn't a long-term view, really. They didn't really understand how the health of the soil translates to the health of the plant. If you are looking at it from a purely chemical viewpoint, the soil is just a medium in which to transfer nutrients to the plant, and then the water transports the water-soluble nutrients to the plant. There's one lady up in Oregon, Dr. Elaine Ingham. The Soil Foodweb is her thing. She's so into it! She's in a microworld all the time. I mean, the closer you look, the more there is to see. There's a whole world that we need to look at. We're just hitting the tip of the iceberg. We don't really know anything compared with what we need to know. There's still a lot of unknown, uncharted stuff that needs to get figured out. But I totally have confidence that it will be figured out, and it can be. We can be, in the long run, probably much more productive than conventional farms, just purely in what we can produce. We're not depleting the soil over time.

California doesn't get any rain for six to eight months out of the year. The three W's of farming, the holy trinity of farming is: work, will, and water. I think it was Wicks, an early 1900s horticulturalist, who coined that. It's true. In California, if you don't have water, you're going to be hard-pressed to grow most items. Water has always been an issue here. Even though I'm along the river, our wells down there are pretty shallow, and there've been times when we're basically pumping those wells dry, and there isn't enough coming into them. I do some dry farming. I do some tomatoes dry-farmed. I have done potatoes and winter squash. But lots of things I grow, you cannot really dry farm them. Lettuce and spinach and cilantro and parsley and things like that need regular water, weekly water.

Now the challenges are more becoming competition. Since the only part of agriculture that is actually growing is the organic industry, a lot of conventional farmers are converting. You see pretty intense competition in our local area, because this is where a high percentage of those growers are concentrated. The Pajaro Valley and the Salinas Valley, that's where a lot of it is happening. There are the natural-world challenges, like weather and water and nature and insects. I can control most of those things pretty well. I've kind of got those down. Once in a while you'll lose a crop planting here and there. But I'm pretty diverse. I grow a lot of different things. I'm planting every week. I have a very, very wide diversity of different plants and flowers. But I think the competition is probably the thing that's going to be my biggest limiting factor.

The farmers' markets are really good for [educating people about organic produce] because you're dealing directly with people. I don't always make it [to market] myself, but at least it's that direct interaction where people can ask you about stuff. You get people coming in new who have just come to a farmers' market for the first time. You have a little more time to talk to people than in a store. It's just a produce worker in a store, whereas at the farmers' market a lot of people are coming by your stand.

And you're gambling. I borrow $150,000. I go into debt $150,000 every spring. I spend a million dollars and I make a million dollars. That's what I tell everyone. That's what it costs to keep going. If you're spending that money and not making any money, you're going out of business pretty quick. Salaries, fertilizers, seed, insurance, containers. Containers are the number-two expense next to labor. Boxes, cartons, and twist ties. That's where the farmers' markets [can help], recyclable containers. You can take stuff in a plastic tote instead of a box, or you can use that box four or five times. That was one of the things that the direct-marketing regulations allowed. You can bring things in nonstandard containers, whereas before you had these marketing orders where you had to use a brand-new container, and it had to be a certain waxed box this size. I think it was probably health regulations. And yes, standardization was a big thing because of the way things were marketed on this mass-market scale. You want to know what you're getting when you order a case of something so it's predictable.

I've gone to a 100 percent biodiesel. All my tractors are a 100 percent B-100. My big trucks, my International ones that run to San Juan Bautista, are running on it too. Pacific Biodiesel. They deliver. I get three hundred gallons at a time. They bring it to the farm. And it's actually cheaper, because for the farm use you get off-road credit. You don't pay road tax. I'm paying less than petrodiesel. And it's better for your engine. There's no carbon buildup. In the long run, it saves you maintenance on your vehicles. It's a good deal all the way around. Sometimes [the fuel is made of] canola oil. Sometimes it's soybeans, safflower, grape.

This whole ethanol thing is a joke. If you're using corn for ethanol, it is such an incredible waste. It is the worst energy conversion crop you can imagine. You put one unit of energy in and you get 1.5 units of energy out. Switchgrass, something that is really easy to grow and has no pests, is at eight to one. Corn farmers are looking for subsidies. It's all about being able to use it for different things, too. Because it's used for corn syrup, animal feed. This whole pyramid of mechanized agriculture has developed to feed this system. It's not the most efficient, best thing for the planet to do, but it's hard to change gears. And you've got these interests that are lobbying against changing gears.

The guy that invented the diesel engine foresaw farmers having a fifth of their land dedicated to growing crops for diesel fuel. But he got thrown off the boat. And then the petrodiesel people came on board. It's just like pesticides. Pesticides weren't around before World War II. All of a sudden, all this technology that was developed for warfare had to be used, because you had these people with these interests pulling the strings in the government. Top-down thing, you know. Same thing. Same exact thing.

Our agriculture is in a horrendous situation right now. I really don't know what's going to happen. I'm worried about shortages because of the fuel thing and all the regulations. Are we just going to be left with these huge corporate farms? It's really difficult for a small farming operation to operate. There's not a lot of incentive to pass on a small organic farm to somebody. They really have to have it in them to want to do it. It's not a cut-and-dried thing. I've been lucky. I do have these pretty killer locations. It's my life, for sure. It's been something that I do because I love to do it. You can't be what I consider to be an organic farmer and just do it for the money. Because you're not going to be doing it right, for one thing. You have to immerse yourself in it. You have to become part of that web.

Ken Kimes and Sandra Ward

Photo: Ellen Farmer

Ken Kimes and Sandra Ward both grew up in Southern California. They met in the Los Angeles area but moved to Santa Cruz in 1980. Together they founded New Natives Farm, a greenhouse-based farm certified by California Certified Organic Farmers (CCOF) in 1983 and located in Corralitos, California. There they tend organic sprouts, including micro greens, wheat grass, pea shoots, sunflower sprouts, broccoli, and beans. In addition to managing their farm full time, Kimes and Ward are both outspoken activists who have dedicated more than thirty years to the sustainable agriculture movement. In 2010, after the date of this interview, Ken Kimes had a serious farming machinery accident, losing his right hand and injuring both of his arms. Ken and Sandra recieved a tremendous amount of support from the community, which organized several benefit farm dinners for them, among other fundraisers. Sandra says, "Many, many times we were stunned by the awesomeness of the caring, concern, generosity, and authentic open hearts that surrounded us."

Kimes: I was born in Southern California. My father sold farm equipment. I spent my formative years in Riverside, from about the age of six on. Sandra and I met in Riverside, and we moved up here together to Santa Cruz County.

Ward: I grew up on a piece of land in Elsinore, California. I think we had about five acres, and my father planted a bunch of walnuts. When I was in my little bassinet, my mother was picking the neighbor's almonds, and I still remember what that smelled like. We lived in a little mud shack, and before I was five I have lots of memories of being a little kid out there in the sticks, loving every bit of it.

My mother was a healer. She took care of seniors, elderly people. She would blend them up a drink full of dandelions and mustards and give them that, and then she would dye their hair orange. We were all just a very happy family. Very colorful. She was also a redhead. [laughs] In that way, I have followed in my mother's footsteps. She was very proud of me when she found out what I was doing, because that was definitely right up her line.

We moved to Riverside when I was about five. It's kind of interesting that Kenny and I paralleled all that time. I met Kenny when I was in my early twenties in Riverside. We moved up to Santa Cruz in 1980, when the smog was so bad in Riverside that you couldn't see the tops of the palm trees. Shortly after that, I started doing a little bit of growing. Kenny went to work in produce so we got into more of the food side of it.

Kimes: Some friends had moved to Santa Cruz, people we'd known for quite a while. They said, "You guys should come on up here. It's great!" He knew a wheatgrass business was for sale up here. He was working for The Juice Club, which was the badly-managed predecessor to Odwalla. But it did prove that there was huge niche for fresh juice.

We moved up here with the intention of buying that wheatgrass business. But when we expressed interest in buying it, the people who wanted to sell it, who were named Windspring and Rainbow (perfect), decided it was too valuable to sell, so they didn't sell it to us. So we found other work. Eventually they just gave up and went away to Hawaii and we took over the business.

It was evident that we needed more space, so Sandra called about a greenhouse space we saw in the paper for rent. That was our friend Rocky, who was doing houseplants in these two ranges of greenhouses. He was in one range, and he rented us the other. That was real fortuitous, and he was real kind to us because we didn't know what the *heck* we were doing. We didn't have a *clue*! [laughs] We had absolutely no idea what we were up to. We were just winging it in the most amazing way. I can't even tell you how little we knew about what we were doing. So we moved into that greenhouse and started working pretty successfully, I guess. It was a ton of work and the returns were not particularly good. We kept that greenhouse going for about six years, and then Rocky wanted it back to grow plants in. So we needed to look for another place.

[Initially] we couldn't find a greenhouse space that would work for us, because we wanted to grow organically and all of them were packed up against commercial growers. It was pretty evident that they were going to have a problem with us growing organically. They were as worried about us as we were about them, that we would infest them because we're not killing everything. That we would let anything live was just too much for them to take. We'd go there, and it was like, oh, my goodness. We were afraid to walk, because everything was so killed, you know. How were we going to share water? What if the runoff— Oh, my goodness. Just toxic, you know. How are we going to do this?

About two years [after we started the business], or maybe even less, there was the [medfly] apple pest. They quarantined Santa Clara County. It was a really bizarre time. At night they would fly helicopters over San Jose, dropping malathion on the entire frigging town and populace. They would fly helicopters in formation of eight, and they would fly a grid over San Jose dropping malathion on everybody. Your car would get splattered.

The state agricultural commissioner had threatened to quarantine Santa Cruz County if we came up with medfly in the apples. So we're working in the greenhouse one day and all of a sudden this helicopter shows up across the street. There're people out there picking apples, and this guy is spraying the crap out of a helicopter all over this orchard. So we closed up the greenhouse as quick as we could, and called the ag commissioner, and of course they didn't get around to it for a week or two. And when they came out they said, "Well, we called up the applicator and he said it didn't happen." They were trying to fight the medfly off, and trying to fight off the quarantine, so this guy was doing everybody a service by spraying all the apples in the Pajaro Valley for five bucks an acre.

When the ag commisioner shows up and tells you that what you saw didn't happen because the applicator said it didn't happen, you go, "I need some pals." [laughs] We need some cohorts in this thing. I had been hearing about this group called the California Certified Organic Farmers, so I thought, you know, if we're going to farm organically we're going to need some buddies. So we called them up, and they said, "Well, yes. We're going to have a meeting. You should come by." There were about, I don't know, ten or eleven people? It was in Watsonville at the ag extension office. They were all very nice and really welcoming. They were glad to see somebody show up, and I think we were the youngest ones in the bunch. Which was interesting, because it's so changed now. I mean, it's such a young person's up-and-coming business. But at that time, we were the youngest ones there. We were, what, thirty-something?

They had their little inspection regime, and they had a set of standards that were just barely codified in law. It was the most rudimentary law in the world. There was no way to enforce the law. It was under the health department. It wasn't under the ag department. The health department didn't want to know about it. They were like, please, don't even call us on this issue.

So we went to the meeting, and then they came out and inspected. Then Barney Bricmont started to convene meetings around trying to get a new [organic] law passed. We were trying to pull together a meeting with people on the North Coast, because there was a North Coast and there was the Central Coast [chapter] of the CCOF. They weren't all that in line with each other about who was going to use what and what the actual standards really were. So Barney would have meetings at the Live Oak School. People would drive down from up north and we'd try to hash out these issues. A lot of what you find in the 1990 California Organic Food Act was worked out by Barney and the gang over at Live Oak School, the basic kind of standards.

I remember an unusual meeting at the Extension Office in Watsonville. The University of California at Davis was talking about putting some money into research in sustainable or organic agriculture. They were going around and doing these listening sessions as to whether or not farmers wanted to do that. Well, the meeting in Watsonville was packed with a bunch of conventional growers who didn't want anything to do with this. They didn't want any money spent on it. I think Sandra and I were the only organic farmers there and tried to stick up for organic farming that we knew almost nothing about. I remember these old guys. They'd say things like, "Yes! I got two rows of strawberries at the edge of my field. I never spray those or do nothing, and they're full of bugs. That's what organic is! Full of bugs."

Ward: And organic is "lazy." "They're only organic because they don't want to pull the weeds and clean it up." That was their idea. That was what organic was all about, do-nothing farming.

Kimes: Yes, a bunch of hippies. [The apple farmers] were really concerned that the organic farmers were going to spread bugs all over the valley, because they weren't spraying like they needed to. There was just a load of crap to wade through. There's a lot of prejudice that still exists today. There's a huge push-back, a really big huge push-back.

I remember I got into an argument with one of the farmers there. He was a real adamant user of chemicals and a real believer in chemical farming. After the meeting we got into an argument out in the hallway. All these meetings are contentious. He is going on, "You can't farm that way. It's not going to work. There is no way anybody could farm without these chemicals. These are the product of human intelligence and research, and we've come to this point, and why go backwards in human history?" A lot of these chemicals, of course, are made from petroleum originally. I said, "Well, what are you going to do when we have another oil embargo and you can't get these chemicals?" And he looks at me, and he takes about two seconds and he goes, "We'll just dominate them." [laughs] I'm like, well, okay. Not much I can say to that. And you know what? It came true, didn't it? It's all about dominating them at this point. It's all about getting the oil out of the Middle East, no matter what it takes. We're going to do that. So he's right. He had it. Twenty years ago he had figured out exactly how we were going to conduct our foreign policy.

Ward: He came over here once to visit. We showed him our soil, which was just beautiful! I turned it over for him and it was just amazing, amazing. And he jumped out of his skin. You could see it. He goes, "Oh, my God! You gotta kill that." Because of all the organisms. You know, we had worms. Things you couldn't even see. But he could sense that that [soil] was full of life. And it scared him to death. It scared him. He was just absolutely terrified. He said, "The first thing I'd have to do is sterilize this place." We got a lot of that early on.

Kimes: There was a lot of animosity early on about organic farming. The conventional farmers had the Farm Bureau. They had the Watsonville City Council. They basically were very comfortable and confident and assured in their position. For them, organic farming probably looked like the whole cultural revolution that came out of the sixties and seventies, a bunch of hippie kids and this foolish stuff going on at UC Santa Cruz.

If you read any of the ag industry press, it's all filled with advertisements for chemical farming. It comes out of the fifties and the post-World War II era, I think, which is that in these modern times we have modern science at our disposal, and we should be using modern methods to do this. They did see better yields. They saw less pest pressure, at least initially, and more return for their effort, initially. I think a fair amount of that return has leveled out and dropped off relative to organic agriculture. But that's where it comes from. It comes from that "the future is so bright we've gotta wear shades" kind of thing. Science is going to solve all of our problems.

And I think it's important to understand, especially now, that the research from universities is paid for by the chemical companies and the ag providers, so to speak. So they're not going to tell you something different. And you can't get money (until very recently, and it's still a very small piece of the pie) to do organic research. I mean, what do we need, right? We aren't going to buy a whole bunch of crap. So who is going to pay for the research to tell us to buy a bunch of crap? There's more stuff on the market now for organic farmers to buy than there ever has been. But initially there wasn't diddly. There wasn't anything to buy. They weren't going to do research around nothing. The university, unfortunately (and I'm sure they would contest this) has been sort of hand-in-hand with the conventional chemical industry, or the industry that provides chemicals to conventional ag.

But I got to say, the organic movement never attacked conventional agriculture. We just called it "conventional practices." We didn't say "nozzle-head practices," or "spray maniacs," or anything like that. We just said, "We're organic farmers and that's a conventional farmer." We really didn't go out of our way to do any gratuitous attacks on these people. We weren't a particularly militant bunch. We were just trying to survive. We never felt like we needed to go out and protest, or march, or put up signs, or anything. We were just a lonely bunch of people that were like, [whisper] "Do you think that way?" "Yes, I think that way." "Okay, let's meet." [laughs]

We weren't particularly ideologues. Most of the people that we encountered were basically libertarians. They were like, "You do your thing, I do my thing. Let's not get in each other's face about it. I'm happy farming without chemicals. Go farm with chemicals. I care, but I'm not going to try to stop you from doing that." That's why it was a puzzle for me that conventional farmers would want to stop us from farming organic.

There is a positive thing about working against that kind of adversity in that, boy, you get your argument chops down really fast. You got to get real solid in what you believe. You've got to develop your philosophy, and you got to say, okay, I'm here, and I'm going to follow through with this. Because obviously there was no real market. We were actually being hard on ourselves beyond what we needed to, to sell to the organic market, because it was not well defined, what it took to grow organic sprouts. It was a time of not very many of us. In fact, we were the only certified organic sprout growers in friggin' America, I think.

Ward: Politically, our personal experience was that [health regulators] were very, very prejudiced about sprout growers. In fact, our first meeting in Northern California at the California Department of Health, we didn't know a lot about what was going on, except we had been called to the table. They actually locked us in. They did a PowerPoint presentation, but before they got started, they had one of those big pads that they'd draw on. And the first picture, they had a dinosaur eating sprouts. They said, "You guys are out of here." That was their, "Hello. How are you?"

The main recommendation from the FDA is to do this 20,000 parts per million chlorine soak on your seeds. That concentration is about one cup of chlorine to one cup of water. I don't know of any concentration in that amount anywhere else in the food industry. They set that up as an emergency that has just been reinstated for the past ten years. I don't think there's any test on it. I find it really dangerous. We stopped growing [alfalfa and clover sprouts] because that amount of concentration would, in our very, very small operation, mean that we would be using what would be comparable to about 380 gallons of chlorine a week. When you are looking at a substance that really does create cancer, then I couldn't do that, not for myself, not for my employees, not for the water. That recommendation was so punitive, not backed by science.

It was absolutely so prejudiced and so biased. The agenda was already set. There's no doubt about it. Of all of those sprouters in that room from Northern California (there were probably about forty of them), after about three years only three of them were left. There were even growers that followed everything, all the over-the-top testing that was recommended, and all of the chlorine, and [the FDA] would come in and still put them out of business. There was nothing you could do. So we said, "You know what, we're going to just stop this right now. Alfalfa and clover are great, but this is going to be our business, so we need to stop."

Kimes: People's hands were being burned. There is actually a disease that comes from being around too much chlorine. You break out. Miscarriages go up. Respiratory problems are huge. Chlorine floats in the air and it actually attacks metal. So it rots all the light fixtures and the metal fixtures in the plant.

The Department of Pesticide Regulation came to one of the sprout gatherings that the health department put together and told us that this solution of chlorine and water that we had after we soaked the seeds is a toxic material, and we need to hire a toxic material handler to deal with it. And the Department of Health Services was like, "Oh, don't tell them that." They didn't want it to be thought that they were recommending toxic levels of a chemical.

It was a real problem for us to give up growing alfalfa and clover, because it was 30 percent of our income. It really set us back. But it was the right thing to do because it took the target off our chest. They were really harassing the alfalfa sprout growers.

There is a huge cultural problem. There is a culture clash. The way I illustrate that is that we've been inspected by the FDA probably a half a dozen times now. The FDA is intimidating because they have police powers and everything else. They are very serious people when they inspect your farm, for the most part. You never know when they are going to drive down the driveway. And they will take up a day or two in the process of doing it. You have to drop everything and spend all your time with them. You can't do anything else when they're here.

The first FDA inspector we had in here told us during the course of the inspection that she never lets her kids eat fresh produce or fresh fruit unless it has been parboiled and peeled first; they never eat at a salad bar; and they generally don't eat fresh produce at all because she considers it unsafe. This is the person who shows up with the police powers to inspect a small farm who is trying to enhance all those values of: natural environment, eating fresh foods, a good variety in your diet, harvesting things when they're ripe and ready to eat, and all those kinds of values that a big segment of us have come to embrace. But this person has the police powers that run—her basic personal philosophy runs counter to all those values. She told us that, and then we said, "It's time for lunch. We have to eat lunch." And we were determined we were not going to feed this person.

Ward: I would have invited her. I knew she wouldn't eat it. I would have totally invited her.

Kimes: And then she says, "Okay. Well, where's the nearest McDonald's so I can go eat?" We are not communicating, really, across that barrier. I mean, talk about a culture gap. This woman came in and inspected us as if we were a slaughterhouse, or a meat-packing facility, or a food-producing factory, rather than a farm. She inspected us with all those same standards. And it was just like, what are you talking about? It was like a person from another planet. Or we were another planet, as far as she was concerned. It was so bizarre.

You can't believe the level of stress [this regulatory approach] adds to your life, because as a farmer you're thinking, I'm producing great food, and I'm feeding people and nourishing people. You're trying to build up your soil and different things like that. And here come these people who say, "You're dangerous. Prove to me that you're *not* dangerous." That's the assumption. That's the beginning assumption, that this food is guilty of danger to the public until you prove to me that it's not.

Ward: Even before you can have a chance to prove that it's not, you have a story in the papers, in many papers, instantly. Even if you do prove it, that same story probably will continue to run. I could see it so easily shaping itself to say that anyone who is not using chemicals or sterilizers or whatever, is being irresponsible.

Kimes: The product is guilty until proven innocent, and that's becoming the starting point for all of this food safety stuff. You are producing dangerous food because it's produced outside in a field with the natural environment, and you've got to prove to me that you are guaranteeing the safety of this food when it comes to the consumer.

Jim Cochran

Jim Cochran founded the first modern commercial organic strawberry farm in California. His company, Swanton Berry Farm, is also famous as the first certified organic farm in the United States to sign a labor contract with the United Farm Workers (UFW). Swanton Berry Farm offers their workers low cost housing on site, health insurance, vacation and holiday pay, a pension, and other benefits, including an employee stock ownership program. Cochran has become a resource for other organic strawberry growers, and in 2002, the U.S. Environmental Protection Agency awarded him the Stratospheric Ozone Protection Award for developing organic methods of growing strawberries that did not rely on the soil fumigant methyl bromide. A key component of Cochran's success was his partnership with UC Santa Cruz agroecologists Steve Gliessman and Sean Swezey in on-farm research. Travelers along the North Coast of Santa Cruz County visit the Swanton farm stand on Highway One, where they pick strawberries by the sea, and savor the jams and other treats concocted in the kitchen. When no one is minding the store, customers pay on the honor system, a lesson in trust that Cochran encourages.

I came to UC Santa Cruz in the fall of '68. My degree was in child development, a psychology degree, but I really got that by default because I was teaching nursery school at the time and it seemed appropriate. My real interest was nineteenth-century European intellectual history. I was in Merrill [College] in its first class, and we were located right next to Alan Chadwick's garden. So even though I didn't really know him personally, I was friends with a number of the people who worked there, and of course you wander over there during breaks and so forth.

While I was at UC Santa Cruz, at Merrill College, I was living with a bunch of Chicano students. We'd all gotten a big house someplace and crammed ourselves into it. One of the students told me about a job opening. They had hoped to find a Chicano, a Mexican-American student to fill this job, but none of them really wanted it; they wanted to pursue their education. And they said, "Well, you know, Jim can do it" (because I spoke Spanish), and so I got a job as an assistant to the people who were organizing a farmworker-owned production co-op in Watsonville called Co-op Campesina.[1] It had, at that time, four members. They had leased four acres of land down in Harkins Slough in the Watsonville area and were growing squash. My job in that enterprise was not central, in the sense that I was not one of the organizers or one of the members, but I was the "gofer," the guy that checked things out and ran around and did everything. I did that for five or six months. Then I worked at the migrant labor camp as a teacher's aide in the childcare center for about six months.

[About] 1970, I inherited $12,000 from my grandfather and decided to buy a piece of land in Santa Cruz County. That was the smartest thing I ever did. It was remote, up here in Davenport area, up in the hills here, and it was covered with brush and stuff, so I spent several years clearing brush and building a little tiny cabin. I got a job at the university children's center and worked there. I spent four or five or six years finishing my degree, working at the childcare center and building a house. Then I worked for a couple of years helping a group of weavers organize into a wholesale operation.

My friends who had been working on co-ops in Watsonville had been doing it continuously since the time that I left. The co-ops had grown and become quite a big enterprise, with a couple of hundred families and a thousand acres. There were seven or eight co-ops covering from Watsonville down to Santa Maria area. However, that was also at the time when interest rates were 18 1/2 percent, and many of the co-ops had expanded too fast. Some of the land purchases turned out to have been a good thing, but in most of the cases, they had taken

---

[1] For more on Co-op Campesina see an article by Peter Barnes written for *The New Republic magazine* in 1971-72 available in full text at http://www.progress.org/barnes11.htm.

on too much debt and expanded beyond their capacity to produce a
high-quality product efficiently. It was fairly typical of small businesses
everywhere. A lot of them expand too quickly. They were in trouble
with the banks. They needed somebody to help sort out their loans and
how to survive. It was all extremely complicated. I was recruited back in
because they knew me, and because I could talk to both farm workers
and bankers. I somehow was able to cross both of those worlds. And I
was willing to work eighty hours a week for not a whole lot of money.
So I did that for about four years.

[At the co-ops we] used a lot of chemical pesticides because the
members all loved pesticides and loved spraying poison on things.
But I had come from Santa Cruz and had a little bit of exposure to
organic farming, and I kept asking people in the business—from the
members in the co-op, to the Strawberry Commission, to people who
worked in Salinas—I said, "Can't we grow a strawberry that tastes
good, and can't we grow a strawberry that doesn't have pesticides
on it?" And the answers to those questions were, "No. What people
really want is a big red strawberry, and they don't really care what it
tastes like, and they don't really care whether it has pesticides on it."

I spent about two years trying to talk people into a different approach. *Organic vs. Convention. Experiment*
Finally another co-op manager and myself rented four acres of land
down here in Swanton Valley. We did half and half, half chemical and
half organic, because the chemical industry did a pretty good job in
scaring the daylights out of us. I have to admit I was a little bit less worried about what would happen than my buddy, the guy who was my
original partner there. So we did side-by-sides. It turned out that our
[organic] production was a little bit lower, and we had more problems
when we farmed organically, but it wasn't disastrous.

We started selling direct to the public, running around selling
to restaurants and just right in the immediate area here. We made
enough money to pay all of our bills but actually not enough money
to pay ourselves. For the first three or four years we seemed to be
building our reputation, but if you looked at what we were getting
paid for the product, we just weren't getting paid enough. My friend
was going to MBA school at Santa Clara, my partner, and he said,
"Well, I can do the math. It isn't going to work out." He decided to
get married and finish his MBA and go on and do something else.

So that left me. By then I had a house, so I put my house up as col-
lateral and borrowed money and kept going, and kept working on getting

the customers to pay more for the product, because it was abundantly clear that I just could not afford to sell organic strawberries at a 10 or 20 percent premium to conventionally-grown strawberries. My yields were so much lower and [labor] costs were so much higher. We hired people to work with us, and it wound up costing way more than what the conventional berries did. It took about five or six years to move the price up to the point where it needed to be, which is essentially about double what the chemical strawberry price was. At that point, I started to feel a lot more confident about the future. I was able to stop working during the winters. I [had] worked as a laborer on construction projects, and I worked at the nursery school that I used to work at during the winters to make some extra money. After five or six years, I was able to actually make my living off of the farm.

By about 1990 or so—I had by that time been in business for six or seven years—Whole Foods came along, so I started selling to Whole Foods. The fact that they were willing to pay more [for organic] and buy a lot helped me to be able to expand. I was the only organic strawberry grower at the time. After a few years other people started in, but by that time I had established myself as a reliable supplier of high-quality product. So for a while I pretty much had that to myself. As the years went by, more and more people started getting into it. Every once in a while somebody would come in and sell product at a price that I knew was too low but they didn't really know that. There's a learning curve, and people tend to be a little too optimistic about what it costs them to grow something, not really adding in all of the costs. To this day, there are a lot of people coming in and undercutting you in price. They last a couple of years and then they drop off and go away. But at this point, there are a couple dozen, at least, organic strawberry growers in the area.

I got [on-farm research] help from Steve Gliessman up at the [UC Santa Cruz] Agroecology Program. He was instrumental in providing the scientific underpinnings for what I was trying to do. Also encouraging, in the sense that when you're out there all by yourself, it's nice to have somebody come along and say, "Well, you know, maybe we should try this." Being sort of hopeful, you know? Because at that time, in the very first years, I think a couple of people had grown organic strawberries when there was a one-year waiting period to become organic. If you took over a parcel which somebody else planted and established, using fumigation and chemicals and so forth, and then waited a year, then that same parcel or those same plants would produce "organic" strawberries. A couple people did that a couple times, so there would be an influx of fruit that had been taken over. But that dropped off after a couple years, especially after the waiting period [for organic certification] became three years.

My goal was to become primarily an organic strawberry grower and to start from scratch—to plant the field in strawberries and carry it all the way through, plant them into an organic field. I went back to good, old-fashioned farming practices, which are crop rotation and adding good soil amendments. Managing soil disease was a problem that Steve Gliessman and I worked out, looking at some of the old literature and coming up with the notion that the Brassica families—plants like broccoli or Brussels sprouts or mustards—might help suppress soil disease. And sure enough, when the farm grew a little bit, and I leased a piece of land that had been half in Brussels sprouts and half in artichokes, and I disked the whole field and planted in a section that had been part artichokes and part Brussels sprouts, the part that had had Brussels sprouts had less soil disease. That really accelerated the practical implementation of the theory, because I said, "Gee, not only is it theoretically possible and there's some references in the old literature, but in my practical experience here, it's also true." So bingo, I added broccoli to the rotation and started growing broccoli and cauliflower as part of the rotation, and then adding high-grade compost and various kelp extracts and that sort of thing to the field to try to improve the general health of the soil.

[In chemical strawberry growing] chemical fumigation serves several purposes, probably the most important of which is to kill soil disease down to about twelve inches. They put the plastic over the field. It's the same chemical that they use in fumigating houses for termites. It's deadly and very effective. It also kills weed seeds and gophers and snails and all the little crawly things in the soil, sow bugs—all the kinds of things that could be problematic to a crop. The fact that it kills weeds is a very big deal. It really lowers the costs of growing the strawberries. And then all the things that are in the soil that die because of the fumigation decompose quickly, and they become readily available as nutrients for the plants, so the plants get this extra boost from all this other, decomposed plant life and animal life in the soil. So it produces this terrific, this 20 percent, at least, boost in yield, not to mention the fact that you don't have any soil diseases or weeds. Methyl bromide is the primary chemical. Chloropicrin is another one that they use. Everybody was using this and able to generate really high yields, so they were able to produce 50 percent more strawberries than we [organic growers] were, just right off the bat, because they're using these chemicals. They were also using slow-release fertilizers, which are perfect because you know how much nitrogen you need in March or April, and if you have a slow-release, capsulated source of nutrients, you can meter that out exactly when the plants want it.

If you're farming organically, you don't really know when things are going to be available, and so you're not really able to have as much "control" over the crop. It was really tough. But that's part of the reason that I undertook the challenges, because of the barriers to entry. It's not hard to grow organic peas or broccoli or some crops, so a lot of people planted that stuff and grew it, and there really aren't very many problems that you have with those crops. But there is a lot that can go wrong with strawberries. I correctly assessed the situation in thinking that I would probably have five or six years to develop the system without having a whole bunch of people switch over right away, because it's not that easy to do. That was the case.

It all came together [after the] Alar event in 1989.[2] [There was more demand for and interest in organic food.] CCOF [California Certified Organic Farmers] was doing more, had a much bigger presence in the media. Having Bob Scowcroft and Mark Lipson to talk about organics intelligently was a really important thing, because it widened the market for us.[3] And all of a sudden there were more people, more scientists interested in organic. The Strawberry Commission wasn't at all interested in this subject, but there were stray scientists coming in from here and there. But probably the most important is the peer support that you get from a group like CCOF, where you've got all these different farmers. We're all struggling with the same thing—different crops, trying to get customers to understand what it is we're doing. So it was quite a good time. There was a nucleus of people here, the symbiosis of all of our various efforts.

---

[2] "A watershed event in the consumer food safety crisis was the release of Intolerable Risk: Pesticides in our Children's Food by the Natural Resources Defense Council (NRDC) (Sewell et al., 1989) By NRDC estimates, Alar posed a particular risk to infants and children, who because of their lowbody weights absorb disproportionate amounts of residues from apples and apple products. A *60 Minutes* broadcast and other media coverage about the report created nationwide panic." M. Elaine Auld, "Food Risk Communication: Lessons from the Alar Controversy *Health Education Research* Volume 5, No. 4, 1990. pp. 535-543.

[3] See the excerpts of oral histories with Bob Scowcroft and Mark Lipson in this book.

There were a lot of really smart, hard-working, innovative people, both in the area and then within the larger organization. I think it was important for us to feel like we were part of a larger movement. UC Santa Cruz also provided smart, creative people who were thinking differently, organizationally, about the way the world is organized, as well as how our food is produced. And there were a few co-ops around that were willing to try our product. They, being fledgling organizations, were willing to try different ways of selling products.

I got my start in the farming business from what is now called the social justice angle. I was interested in finding a way for farm workers to get a better deal in the system. The co-op model captured my imagination. So what we were trying to do was to translate that into a farm worker production model, which is really different from the model that was done in the Midwest, where you had established farmers who owned a farm and a car, and they had been educated, and they had equipment and so forth. They had a business with substantial assets, each one of them, and they got together and marketed their product together. Well, this was working with a group of people who had little or no education, little or no understanding of the larger economic culture here in the United States, little or no understanding of the crops that were being grown. They knew how to grow beans and corn and a few other things (some of them), but not complicated crops like strawberries or other crops. So it was really quite a challenge. The idea when I first started out was to just survive on my own with a handful of employees. It took ten years to at least feel like I was just surviving. Even longer, really, maybe more like ten or twelve or fifteen.

Then I got to thinking. Making a commitment to being an organic farmer and being certified meant that you went through a process of certification, and you followed a set of rules that not only helped you grow and learn about farming and follow a system, an established system, or an evolving system, but had the effect of an outside agency or person coming in and verifying that you were in fact following those practices. So it was a dual thing of developing good practices and having those practices reviewed. I thought, gee, if we can do it with fertilizer and plants and soil, why can't we do that with people?

There wasn't much interest in the organic community in certifying labor standards. So I thought, maybe the thing to do is to become involved with the UFW [United Farm Workers union], which has, of course, had a long history of working for farm worker rights. So we

[Margin annotations: Social Justice / Coop / & / UFW / labor as part of organic package]

got together early in '98 and negotiated a contract. It was a complex process with all sorts of things in there that I'd never thought of, but it was an excellent learning process for me. Labor issues [were] not being addressed in any formal way by the other farms, and I felt like it was really important to do that.

We've now had a contract for eight years. An employee has a problem with something, they talk to so-and-so and so-and-so, we have a little meeting, and we resolve it this way. There's a process for talking about wages, and we have a health and dental plan, and we have vacation pay, and we have holiday pay and a pension plan. But it's expensive, so our costs are probably 20 percent higher than other farms' costs when you add all those things in, not to mention the cost of the extra effort to the human relations things.

[We've] started an employee stock ownership plan, which is also not done in agriculture. We're the first farm in the country to attempt it. We added a profit sharing program on top of the wages, so when we have a profit we contribute a pretty significant portion of the profit to this profit sharing plan just for employees. And so in a way, you can't really "buy" the stock. You have to earn it. It's all done in a democratic way. It operates under the ERISA [Employee Retirement Income Security Act] laws, which are Department of Labor laws. They're very highly regulated. It costs us twenty-five thousand dollars a year just to administer it, between lawyer's fees and accounting fees just to administer it and be in compliance with the law. It's very expensive. But it's a great system, because it's a way for people to accumulate equity in the company and then leave in an orderly manner because the company buys the stock back from them. It's a system that's been worked out over the last twenty-five years, so there's this whole body of knowledge about how the agreements are written up, and what kind of training programs people can have, and what's the fairest way to allocate the stock, what's the fairest way to value the stock.

You know, first it was organic. Then it was just basic workplace issues, and then it's this ownership piece, which is another level of complexity that we're just embarking on, so it's going to be a few more years before people have much of a grasp of what it's about. But we've had our meetings, and they get a certificate with how many shares of stock they have, the value. We make a cash contribution sometimes, too. So they have, like, their own IRA account, and unlike most IRA accounts, they have some control over the value of their stock. Because they think about things and say, "Hey, let's try this" or, "Let's quit doing that." They have the ability to make things work.

I think there's a lot of interest [in our programs] among a number of small- or mid-size farmers, mostly the mid-size people. The people that have ten employees or more, or eight employees or more, are saying, "Gee, how do you do this?" and "How does this work out?" and "How much does that cost?" I don't think there's a lot of interest in signing a union contract, but there's a lot of interest in trying to do many of the things that the union contract stipulates. The truth is that very few farm workers in this country are represented by a union, just a tiny, tiny percentage.

We have a real stable work force, and we tend to have a higher percentage of older workers, because the health care is more important to them. They have a hard time keeping up with the twenty-two yearolds, and so getting a flat wage is an appealing thing to them. They don't have to worry about getting fired because they're old.

The vision for the future of Swanton Berry Farm is going to come not just from me. I'm trying to institutionalize something that is an organism. I don't think about the farming as much now as I think about trying to create an organism which farms, and houses people, and educates people, and feeds people. It isn't about me. It gets beyond the personality of the founder, or the individual key personality. There're lots of personalities that come in and do this. I have other people who speak to the public about the farm. I'm not the sole spokesperson or the sole idea person for the farm. I'm trying to get away from that.

It's really tough if you are trying to do all this innovative stuff, to do much more than break even over ten years. We've reached a nice size. We have twenty acres of strawberries. That's less than one acre for every year we've been in business. A lot of farms have grown much faster. They're way bigger in organic strawberries than we are. We only grow 1 percent of the organic strawberries in the state of California now, or 1 1/2 percent. We used to be 100 percent. I've gone for depth rather than expansion.

What people really want is not money, but security. The part of life that is work, it's about fulfillment, but it's also about being able to feel secure about your future. Am I going to have enough money to live on comfortably when I retire? Am I going to have good enough health insurance if I'm sick? Am I going to be able to work here even if I come up with some disability or another? Am I going to be able to work in an environment that is friendly, and I don't have to work so hard that

I kill myself, and that I enjoy? I mean, it's pretty basic. If society had those things already set—if you knew that you had guaranteed health care and guaranteed decent retirement—a lot of pressure would be off people to make a lot of money.

Jim Leap

Photo: Tana Butler

With California family roots reaching back to the 1850s, Jim Leap grew up in California's Central Valley. His father was an insurance agent and anti-racism activist, writing policies for the United Farm Workers at a time when other insurers refused the organization's business. Growing up in the 1960s, Leap was exposed to UFW grape boycotts, Teatro Campesino productions, and other activities connected with the farm worker movement. Leap later founded a successful small farming operation of his own, where he empha-sized sustainable methods. From 1990 until his retirement in 2010, Jim Leap managed the twenty-five-acre farm at UC Santa Cruz, designing crop systems, overseeing production, purchasing and maintaining equipment, teaching apprentices, supervising staff, coordinating field research, writing training manuals, and educating students and visitors about the farm.

My father's brother started a successful insurance company in Mer-ced in the fifties. My father worked for him for a while, and then decided he needed to do his own thing, so he moved to Fresno and started an independent insurance business there. He was very progressive, and he'd witnessed a lot of racism growing up in Merced. He was very aware of the way that people were treated differently based on their skin color, so he became a political activist. And as an independent insurance agent, he was approached by the UFW when that organization started to gather steam. It would have been in the mid-sixties. The UFW was gaining assets, they had buildings and vehicles, they were building infra-structure, and they couldn't find anybody in the Delano area to cover their insurance. Nobody would touch them with a ten-foot pole. It was part of the political movement against them. The insurance business is pretty conservative.

So the UFW approached my father, and he said, sure, he would do it. He ended up writing policies for them and representing them. He became the insurance agent for the UFW. At any rate, the companies that he worked with eventually dropped him. A long story short, he ended up losing his business.

Because of my father's involvement with the UFW, I was exposed to a lot of different people and political concepts related to land reform and labor issues. We went to Teatro Campesino plays,[1] and we stood in picket lines and boycotted grapes, and I got exposed to the whole UFW scene and got to know some of the activists. I became good friends with one of the organizers. That was Bobby de la Cruz, who's now a labor organizer in the L.A. area. He grew up as a migrant farm worker. His mother was Jessie de la Cruz, who, in terms of the hierarchy of the union, was right up there with Dolores Huerta.[2] She was, and still is, really well respected, an amazing woman.

At that time they paid a union organizer five dollars a week. [Bobby] needed to make money in the summer. So he said to me one day, "I've got to make some money, man. I'm going to pick grapes this summer." I picked grapes with him. It was a total eye-opener. We were union pickers, but man, I had no idea what field work was like. I was, what, eighteen years old? When I became aware of the conditions in the field, I realized that something is really wrong. We were picking piece rate. It was 110 degrees, four people on a team, and you get paid by the ton. So you work as fast as you possibly can. It's beyond the kind of stamina that would be required to be a pro athlete. And the heat and the dust and the spiders. It's backbreaking, and just hustle, hustle, hustle, day in and day out. I was blown away by the whole thing.

My father had been involved with an organization called the Westside planning group. At the time, a guy by the name of Sal Gonzales was the director of that organization. They got this idea, a lot like the Agriculture & Land-Based Training Organization [ALBA]. They thought, "Why don't we start a program that teaches and empowers farm workers to farm?"

---

[1] Teatro Campesino (farm workers' theater) is a theatrical troupe founded in 1965 as the cultural arm of the United Farm Workers. The original actors were all farm workers, and El Teatro Campesino enacted events inspired by the lives of their audience.

[2] Jessie de la Cruz was one of the founders of the United Farm Workers [UFW] and one of its first women organizers.

So Bobby and his family signed up. They said, "Yes, let's do it." This organization hired an agronomist and a couple of master's students from Fresno State, and they helped these families rent land, gave them all the tools and the skills and the experience to grow their own crops. And the de la Cruz family rented ten or twenty acres, planted cherry tomatoes, a specialty crop that was relatively unheard of at that time. They made enough money that first year to buy the land they were renting.

This was when I was in high school and I was trying to figure out what to do with my life. My father and I put everything we had into [trying to get George] McGovern elected for president. My dad was the county director of the campaign. We actually won in the city of Fresno, but I was so disheartened by that whole thing. At that time I was also involved with some friends who were involved in the antinukes movement, and I became disheartened with that, too. I started thinking direct action is the way to go. It hit me one evening. I was in a meeting with the "no nukes" crowd, and they were planning a direct action. Some of the more radical activists were going to jump over the fence and get arrested at Diablo Canyon. I didn't want to get arrested, but I wanted to support them, so I was there at the protest as a support person.

But then I thought in this meeting: this all makes sense, but what even makes more sense is to just turn the lights off, cut the power, and make an individual stand. So then I started thinking about the farm workers. My dad was telling me about my friends who were doing really well in this agricultural training program. Based on what I'd seen on the picket lines, I thought, wow, if I really want to make a difference, if I really want to be an activist, the best thing I could do would be to find some land, plant some crops, treat people with respect and basically set an example of a viable option.

I eventually ended up on my [own] little four-acre farm, doing all of the work myself, selling crops direct and local. [It began when] I met some very interesting people that were also volunteering a lot of time in the McGovern campaign, and they were starting a commune. [laughs] They were going to be back-to-the-landers. They said, "Come on. Join us. This is going to be really great. We're going to buy land and grow our own food."

We had money, because we were all working day jobs and saving and pooling money. There were six of us. I was certainly the most engaged, in the field. We bought a tractor. We rented some land from a farm advisor, Don May. He gave us a little bit of guidance. We also worked a lot with Pedro Ilic, a small-farm advisor in Fresno County at that time. Our farm cooperative grew ten acres of sweet red onions. We wholesaled. We sold all our cull onions at the flea market. [laughs] And one of the commune members worked in a grocery store, so we started some grocery store sales, too. [Eventually] the commune kind of fell apart, and in the settlement (we had all of these assets), I said, "Hey, just give me the tractor and the cultivators and the truck." So I went off on my own.

I loved everything about [farming]. It requires every skill set you could possibly imagine. I loved the pumps and the irrigation and the tractors and the cultivators and the growing and the planting of the crop. And then through the sweet red onion project, I got to be friends with a local grower about my age, Yoshimoto Kamine. Yosh's dad, Moto Kamine, along with Yosh, ran a very diversified vegetable operation—not organic, conventional. Moto was one of those [Japanese Americans who was] sent away to the intern[ment] camps. He was working his way back up the ladder and had this really diverse farm. This would have been about 1976. I approached Yosh and said, "Let me just work with you for one season, twelve months, but I want to do everything. I want to pick, pack, irrigate, drive the tractor, do deliveries. I want to be exposed to everything. Just pay me what you pay your workers." So that was pretty incredible; he let me do that.

He also had a welding shop, and they were building farm implements. I learned how to weld. I learned how to drive tractors. I learned how to irrigate. I did deliveries. I learned how to tie down loads. I learned all this stuff, a lot of it from the farm workers themselves. He had a solid crew of guys that worked with him year round. I spent a lot of time with them, and they had a lot of fun teaching me Spanish. I had already had six years of Spanish in school, but I really got my conversational Spanish down and learned a lot of amazing stuff.

Every one of [these workers was Spanish-speaking]. And illegal. I witnessed their constant struggle with the Immigration and Naturalization Service. The INS was very active at that time and there were a lot of political games being played. The Kamines were contracting with a grower/shipper. On one large field of red onions they hired a labor contractor for the harvest operation. When it came time to pay the field workers, the contractor called the INS, and many of the field workers got picked up. In the fall, if some of the field workers were ready to go back to Mexico, and the INS came, they'd say,

"Take me home." The ones that wanted to stay and do a little more work, they'd run off into the vines and come back two hours later.

I witnessed so much, I was even more convinced that I wanted to have my own farm. After I was done with my one year with Kamine, I went to work for three years as a field manager for KorMal Company, a large fresh market onion grower/shipper, to get even more experience. We had hundreds of acres of onions, and we did cherry tomatoes, winter squash, lots of eggplant. So I learned the whole eggplant thing, again, conventionally. In the wintertime, I helped weld onion graders and do repairs. I was learning mechanics and all of those critical skills.

My partner at that time, who was in the commune with me, her parents had this piece of property, so we had this little quarter acre that we planted out. This would have been about 1978. That was around the time that Wilson Riles, who was superintendent of schools, and Jerry Brown got together and said, "Let's create a way for schools to have access to fresh produce." That evolved into the establishment of the California Certified Farmers' Markets. Nineteen seventy seven was the first year that farmers' markets were actually allowed in California. So I was growing this produce on this quarter acre at a time when there was not a "certified" farmers market anywhere in the Fresno Area. At that time there was very little easily accessible information about growing practices, especially organic growing practices. Of course, there was no organic certification then either.

I was doing my best [to grow organically] but it was really hard, because there just wasn't any information. All the information I had was conventional. Fortunately, I had a neighbor who was starting out at the same time, who rented ground in my neighborhood at the same time I did: Tom and Denesse Willey of T&D Willey Farms. Tom hooked up with an African-American farmer in our neighborhood, Mr. Po, who had come out from the South and had a little plot of land and was growing mostly beans and greens and potatoes. Tom got a lot of great crop culture information from Mr. Po and would share it with me.

I had another mentor, a gentleman named Ben Franklin, who was an African-American grower out on Jensen Avenue. I met him through the onion deal, because he was growing onions, too. He was just the sweetest, nicest guy. But he was conventional. He wasn't organic. He shared with me everything he knew. In those days, information amongst growers was held very tight, because the markets were so competitive. For example, with the sweet onion deal, everybody had their secrets.

Everybody had a different way of doing things. Ben Franklin took me under his wing and told me that any time I had a question, any time I needed to borrow an implement or anything, he was there for me. He opened his doors completely to me. His neighbor across the street was a good vegetable grower, and Ben Franklin's trick was, to get information from this guy, he would go over there with a bottle of whiskey in the evening, get the guy loosened up a little bit and then ask him questions. [laughs] There was a whole little community.

I had the equipment from the commune. Also, I figured out the whole auction scene. So I was buying farm equipment, because I knew what I had and what I wanted to piece together. I had an interesting relationship with KorMal Company, because they would lease my equipment from me. So I developed a whole system. I had a tractor and cultivators and various tools, and then they would lease it from me, and they also leased my truck from me. It was a good relationship that worked really well for all of us.

I stopped working for KorMal Company at the end of 1979 and started my own farm in earnest in 1980. I rented this beautiful little four-acre piece out east of Fresno on Locan Avenue in '79. It was a pasture, which actually worked out to my advantage because it was so vibrant and healthy. It was heavy ground. I was at the Germain Seed Company on Jensen Avenue, buying seed. I'm starting my own farm, and it's in the springtime, and I see this little flyer for this certified farmers' market starting in Fresno, so I picked it up and I called the guy. It was Richard Erganian who started the Vineyard Farmers Market on the corner of Blackstone and Shaw. I thought, Oh, that's pretty cool. It was his first year. I told him what I wanted to do and he supported me one hundred percent. He had had a couple of growers. In fact, Benny Fouché was one of them, who became a small-farm farm advisor and just recently retired from UCCE [University of California Cooperative Extension].

So we started this little farmers' market in this little parking lot on the corner of Shaw and Blackstone in Fresno. I built my clientele. I planned my farm carefully. Because the farmers'-market venture was total uncertainty, I divided my farm in half. I was going to do half wholesale and half farmers' market. I was very familiar with eggplant production, and I was pretty familiar with bell pepper production, and I knew all of these other crops, like onions, and I was familiar with the mix of crops from my own little quarter-acre project. I knew how to grow squash and tomatoes and many of the warm season crops, so I did close to an acre of eggplant, and a half-acre of bell peppers, and then a whole variety of crops for direct sales. And it was one of those one-in-ten years where eggplant prices were just crazy, so I made a bunch of money on eggplant, and I was selling it to the broker who I had worked with. I felt guilty, you know? I could make $300 a day just picking and packing eggplant by myself. And $300 a day might not sound like a lot of money, but at the time, I was renting my ground for $100 an acre for a year, and I was renting my house for $100 a month on ten acres, so my overhead was not real high.

It was so powerful to grow stuff, load it in a truck, take it to the farmers' market, and then end up with cash in your pocket. I can't think of a more real way to earn a living if you're going to be a player in this economic game. I just can't think of any other way to do it that would be more direct or honest.

So I made good money and I put that right back into implements and equipment, and built a greenhouse. That got me off to a solid start. I don't know what would have happened if I hadn't had that little break. That got me through the winter. I still had to do odd jobs to make ends meet. I pruned grapes in the winter. I drove a forklift for another grower that I knew. I worked construction. I did just about everything. But the farm went really well. My clientele was picking up at the farmers' market. So I had a number of really fun and interesting years there.

Then I ended up renting a house across the street from my little four-acre piece that was on ten acres of bare ground with no irrigation. I planted oat hay on that ground and became an instant oat hay grower, which was a fantastic lesson. I'll never do that again. [laughs] There's no money in oat hay. You have to have a thousand acres to make it work. But it was fun and very educational. I was really having fun with this whole farming thing.

I hired people [to work for me on the farm] from time to time, and it never worked out very well. I was so efficient and so fast at harvesting, I could grade in the field. When the farmers' market really got big, I had to have help at the market handling the cash and keeping the produce stocked. In the summer months, the farmers' market was a madhouse for the first three hours on Saturday morning, so I did hire people to help me with that, but for the field stuff, anybody I'd ever hired, they'd make more money than I would make. I was kind of unique in that way. I did all of the fieldwork myself. I had a bunch of systems down, and I was really efficient at it. I had all the implements and equipment that I needed. It was all mechanized, and I'd plant to moisture. My irrigation systems were really efficient. I had it kind of dialed in.

My wife worked a little bit with me. She helped out when I needed help. She would plant on the transplanter while I drove the tractor. We had a greenhouse and a nursery, and her thing was cut flowers, so she would do the cut flowers and bouquets and dry arrangements. I'd grow the garlic. She'd braid the dried flowers into the garlic and we'd sell wreaths. We sold them at the farmers' market.

I was not organic at that point, but I was using, for the most part, organic practices. All my customers at the farmers' market knew me and trusted me, and I couldn't see going to the trouble of going through the certification. It would have been different if I was a little larger scale or growing wholesale. But that was in the early days of organic, and in Fresno at that time there was not a lot of awareness about organic. In fact, when we did our first little quarter-acre patch, it was organic, and I called it organic, and people kind of steered away from us at the farmers' market, like, "Who are those weirdos?" [laughs] People didn't know enough about [organic in those days]. And the people that did know about it would hang crystals over our produce and see which way the pendulum swung.

There was so much abuse in the organic deal at that time. People that I knew who had access to produce, would just buy stuff, load up a truck, go to the farmers' markets in the Bay Area and call it "organic." The rumor on the streets then was, "Oh, if you just stick an 'organic' sign on it and take it to the Bay Area, you'll sell it out like crazy."

Then I finally figured out the fertility thing with the cover crops. But when you're farming on that kind of scale, cover cropping is expensive, costly. In those days there were some very benign synthetic fertilizers like calcium nitrate. It's nitrogen extracted from the atmosphere with hydroelectric power. It was made in Scandinavia. It was cheap and plentiful and very benign. I had a choice between that for nitrogen, or the fuel costs and the time out of production to do cover crops, and it was a lot of times much easier to go with the synthetic option. But over time, I saw the benefit of what the cover crops do to the soil. It just takes time to work through all of those details and see how it all pans out. My neighbor, Tom Willey, was organic at that time, so I learned a lot from him.

And then, in 1983, it was one of those heavy rain years—it was definitely an El Niño year. It just rained and rained and rained. I was looking at my fields in June. I'm not getting stuff planted because the ground is too wet to work. I opened the newspaper and saw an ad for a job. It was one of those things, you know, when you're looking for something, sometimes it just jumps right out at you. I opened the newspaper and looked at the help wanted, "Wanted: Crop Production Manager for federally funded program to train Native Americans how to farm. You need desired number of years experience in growing vegetable crops." I applied and I got the job.

The money was flowing for that first year, because we had the grant. We set up a community garden in town, and a ten-acre demonstration farm on the edge of town. Native Americans came from many tribes, from the foothills and from out in the valley, and signed up to have a community garden plot or to do the training program.

At the community garden, which was not too far from the co-op extension offices, we had a beautiful piece of ground in an urban setting. We purchased all of the necessary tractors and equipment. We set up community garden plots and trial plots. All my farm advisor friends—Pedro Ilic and another guy who was a weeds specialist, Bill Fischer—saw this as an amazing opportunity. Because of affirmative action, ethnicities had ratings, and Native Americans had a high number, so the farm advisors were very excited to work with me. We did a number of field trials at the community garden site. With Pedro Ilic, we trialed plastic tunnels and row covers and hot caps (all early season production strategies), which he was really into. And with Bill Fischer, we did weed trials, including herbicide trials. I had fun with the whole thing.

So at the end of that one year, when the grant funding had ended, we were folding up shop on that project. Pedro Ilic was there and he was talking to me. He literally grabbed me by my shirt and he said, "Look, if you don't go get a college degree, I'm never talking to you again." He meant it. [laughs] I was like, "Oh." So we talked about it. I asked him, "What do you mean? What are you talking about?" I was thirty years old, [had] graduated from high school, just kind of jumped straight into this farming thing. He said, "Look, you're doing all of this work on your farm. What are you going to do if you break your leg?" I started thinking about it. It started to make sense.

My grandfather was a medical doctor in Merced, and he was always pushing for me to go to college. He had me on the college track ever since I was two years old, which was maybe one of the things that kind of steered me away from college. So I went to my grandfather, who was at that time probably ninety years old, and I said, "Granddad, I'm thinking about getting a college degree." He said, "Yes, let me know how I can help." He covered a lot of my base expenses, and then my mother also helped out with books and tuition as well. I enrolled at Fresno State, kept on farming, and the whole thing worked really well, because Fresno State is on the semester system, fall and spring. Then I had the whole summer to farm.

I studied agricultural science at Fresno State. It took me five years. That would have been 1985 to 1989. I graduated with honors, loved every single class. I was one of the best students in all of the ag classes I took, because I had so much experience. I could relate to everything, and everything made sense.

I learned a lot of non-useful information, but I learned a lot of useful information. I learned the jargon. I learned soil chemistry. I learned plant pathology. I learned entomology. Much of the technical information that I use every single day on this farm here [at UC Santa Cruz] I learned at Fresno State. It's the jargon. It's the lingo. It's what you say and how you say it. I learned all about herbicides and all about conventional farming, which was very helpful. Because if you don't know that, you can't advocate against it, you know?

So I'm just happy as a clam, right? I'm in school, and I'm farming, and the farm is efficient, and I'm actually making money, and I have this financial support from my family to attend college. I graduated. I was really excited about farming after I graduated, because I could see that, boy, now if I have full-time to do this farming, I could do well. I actually made fairly decent money my last year of farming, with not a lot of overhead.

A friend of mine who was the veg-crop technician at Fresno State heard about this job here at UC Santa Cruz, so he brought the job application to me at the farmers' market. He thought about applying for it and he thought, "No, no. This is the perfect job for Jim. I can't do this." So he brings me the application one day at the farmers' market in the fall of '89 and says, "You should apply for this job. This is perfect for you." My response was, "No, no, no. Randy. I just want to farm. I just want to farm." I had everything down and everything was working, and the market was strong, and I had all of these great customers. I could see that I could make a decent living. Or I thought I could.

Randy kept coming back to the farmers' market and asking me if I had submitted the application. "Did you send it in? Did you send it in?" He knew when the deadline was. Finally, at the last minute, he says, "Look, just send it in. You don't have to accept it if you get it, or whatever. Just send it in and see what happens." I said, "Okay, Randy." I filled it out and sent it in.

By now it was January of '90. They called me. They said, "We want to interview you," so I went over to Santa Cruz and interviewed. It looked interesting. I thought, how fun. My mind was fresh with all of this information from college, and what I wanted to do really was continue with college, get a master's degree at Fresno State, and become a farm advisor. That was my long-term goal. That made the most sense for me, because I really enjoyed working with the farm advisors, and I really liked that whole idea of farm research. So this new job opportunity was taking me off track a little bit. They called me back and said, "The job's yours if you want it."

In the meantime, I had met with my accountant who kind of dropped the boom on me. I didn't prepare my own taxes. I don't know if you've ever seen tax returns from a farming operation [laughs], but it's incredibly complicated. I didn't realize it at the time, [but] I had kind of painted myself into a corner. I was a mechanic. I had a neighbor who had a mechanic's shop, who let me use any and every tool. I rebuilt everything. I built a lot of my own implements. Through some tax law I was unaware of, I had to start paying what was called personal property tax, which is really a tax designed for yachts and motor homes and things you don't use much, but that applied, for some reason, to farm implements. So I got hit for this tax for all of my farm implements.

So I said, "Okay, that's fine. I spent fifty dollars in materials to build that bed shaper." They said, "No, what would that cost if you went out and bought it new?" So I got taxed on everything. With this personal property tax, I got hit for all of my implements and tractors. When you purchase a tool or implement or tractor, you can write it off against your tax liability, but that write-off is for either seven or ten years. I'd been farming for exactly ten years. I'd just lost all my write-offs, too.

Before I'd met with my accountant, I was thinking, oh, I can start buying health insurance for my family now, because my income had been steadily increasing. And my accountant said, "No, you're going to be paying quarterly self-employment taxes, plus you don't have any write-offs." From her perspective, my only option was to expand my operation, go into debt, borrow money. And it's like, "No, no, no, no, no."

So all of this is happening concurrent with this amazing job opportunity over here at UC Santa Cruz. And then, to top it all off, my back was killing me, because I was doing this oat hay enterprise, which was insane, and harvesting all my own vegetables, and I hadn't had a vacation since I was in high school.

So this new job opportunity kind of made sense. It was like, "Okay, let's move to Santa Cruz." So I called them back and I said, "Yes, I'll accept the job." They said, "Great. When do you want to start?" or "When can you start?" I said, "Well, how about in September?" (This was in February.) They said, "Well, we need you Monday." I said, "Oh, my God." I had all my crops in the ground for the spring. I was still harvesting crops from the fall and I was doing the farmers' market. I had all my seeds purchased, all my inputs purchased for the whole year. I had this horrible decision to make. I realized I really wanted the job.

Nesh Dhillon

Photo: Tana Butler

Nesh (pronounced "Naysh") Dhillon is operations manager for the Santa Cruz Community Farmers' Markets, which emerged as a redevelopment project after much of downtown Santa Cruz was destroyed in the 1989 Loma Prieta earthquake. The markets have flourished over the ensuing two decades in this agriculturally rich, socially diverse, sometimes politically contentious community. Dhillon's parents both grew up poor—his father in a farming family in northern India, his mother in rural Oregon—but with a preference for fresh, nutritious foods, which they passed on to their son. A Jesuit high school education instilled in the young Dhillon a deep concern for ethical behavior, cooperation, and justice—values that he brings to his work in sustainable agriculture.

I went to a Jesuit high school in Portland, Oregon. They stressed academics and athletics, but they also taught what I thought was probably the most important thing, cooperation. My early educational experiences really laid out everything that I do: thinking and ethics, drawing lines to ethical dilemmas and establishing core values.

I got introduced to the idea of sustainable farming at the University of Oregon. That was my first experience with organic farming, urban farming. They had a program that my girlfriend at the time was really involved in. She was from San Francisco originally, so she was really attuned to what Alice Waters was doing. She got my head wrapped around these ideas of fresh, sustainable, local, seasonal foods—all of these sort of things that as a person who shopped at, say, Safeway, you have no concept of.

Both my father and mother grew up really poor, and my father came from a farming family. The only way that his family would be able to eat was to grow their own food and hunt. So I grew up hunting a lot. We didn't grow our own food, because we lived in Portland, but my parents were always sticklers for buying fresh, quality food. It was really important in the family. We never ate Doritos. I never had soda pop, none of that stuff. So I was lucky. I was weaned on the flavors of wholesome, fresh foods. Plus, I also had a respect and reverence for the harvesting of meat, because I deer-hunted; I duck-hunted; I elk-hunted; I fished. I did it all. So I had a respect for all of those things.

I was lucky because my father grew up in a country that honors food and has long cultural cooking traditions. He's from north India. Food is a huge part of the Indian culture. Even though Western ideals are starting to infiltrate into the subcontinent, it's still a huge part of culture. And my mother came from a Depression-era ethic. For whatever reason, she lucked out, too. They would eat fresh, wholesome food. They didn't have a lot of it, but they would do it.

Interestingly enough, back in the seventies and even probably a little bit in the sixties, when the packaged food started to really hit the shelves, it was considered a luxury item. It was more expensive at that time than the fresh stuff. If you went and bought a bag of Doritos, it actually was much more expensive in real dollar terms in the 1970s. Now it's totally different, because we've manifested this commodity-based agricultural system where the cost of the raw ingredients has come down so much because there're huge surpluses. Everything's flipped around, and that all started basically in the 1970s with Earl Butz and all of the problems with the food price shocks.[1]

In the nineties, I started to make a buying commitment, a shopping commitment. Honestly, this was way before all of the fanfare. There were a lot of things tugging at the inside of me saying, this is too expensive. I didn't have much money to go around. I had that battle for a long time, until I got to the point where I said, it doesn't make any sense to be walking the fence. Consume less. Eat higher quality.

---

[1] Earl Butz served as Secretary of Agriculture from 1971 to 1976. He ran the Department of Agriculture during a period in American history of steep price increases in food.

I understood the ratio of supply and demand and prices and all this stuff. I said, eventually, over time, this product that I'm buying for ten dollars a pound will hopefully be five dollars or three dollars a pound because the supply will increase. And that's actually what's happened, just in a short ten-to-fifteen-year period.

Santa Cruz Community Farmers' Markets started in 1990 in conjunction with the Santa Cruz Downtown Association, after the Loma Prieta Earthquake, as a redevelopment tool. The beautiful thing about it was that it was organized by farmers, and most of them were organic farmers, some of the pioneers in the area. They started the organization with the intent of promoting local food, and less so as a redevelopment tool. But the city wanted something to bring people back into the downtown corridor because everybody was shell-shocked from seeing their downtown disappear [after the quake].

When I came on the scene, which was about 2000, the market was going through a lot of transitions. It had recently broken away from the Downtown Association, and started to operate as a stand-alone non-profit. In 1999 or 2000, I was asked by the new operations manager at the time to work for the market part-time as an assistant manager. My position was to just do the grunt work; cleaning up, making sure there were no dogs in the market, answering questions. It was great, because I was really interested in farmers' markets in general, and I had been shopping at farmers' markets for quite a while. I said, this is kind of an interesting little side job. At that time, I had, like, three jobs. It went up to five at one point. Now it's down to one, thank God.

Not even a year into it, the market ran into a whole bunch of controversy. This was in '01, '02. When the market first came online, all of these people were showing up at the market and playing music, hanging out, drumming, basically having a good time. It was creating tensions. Customers had a problem with it. Vendors had a problem with it. Downtown businesses had a problem with it. But on the flip side, there were customers that loved it and vendors that probably loved it. It was this tough situation. Eventually what happened was the activities that originally were inside the market were pushed out of the market.

Getting the position that I have now was just a lot of chance. At the time the position came available, I was working all these jobs but really wanting to settle on a direction and convinced myself I needed to make some money, so I was interviewing for some very well-paying jobs in the Bay Area. So, I was at this crossroads, about to make a decision about what type of work I was going to do and what my motivations were and I said, what does my heart say? I said, I think I'm going to apply for this position for the farmers' market. That was the best decision I've ever made.

The internal question was, who are you as a person? When you identify what your core values are, you're basically looking inside yourself and saying, what resonates with you? That was one of those pivotal moments, because the other job was definitely not that, but the pay was really good. I was leaving behind a lot of money to do something that called to me on a more intrinsic level.

My duties have changed over the years, partially because I've been able to grow the business. Over the course of seven years, I've gone from being the guy who does it all—picking up all the garbage at the end of the day, I mean, everything—to being more of an overseer, truly an operations manager position. I have a staff of six part-time people now. I knew that when I got the internal functioning of the market to where I wanted it, I needed to start delegating and hiring. I think a really important part of being a business is providing opportunities for other people. I could pay somebody good part-time wages, benefits, and give people an opportunity to work within the farming community.

Farmers' markets have been established in the State of California to support the interests of the small-scale grower, because we have a huge bifurcation within the agricultural community: large, small, there's really nothing in the middle.

We're totally self-sustaining. Our income is derived through stall fees and membership dues. In order to survive, you must bring in adequate revenues, but if the focus is too revenue-oriented, you move away from your mission statement. If you start bringing in more growers, and you create saturation points for some commodities, then smaller growers inherently get hurt. And then what happens? They stop growing the commodity, or they drop out of the market. So you have to be able to keep commodity levels in check. And then the other thing is, some markets over time start to ebb away from being a farmer-oriented farmers' market and become more of a street fair.

Ideally, a farmers' market should only be farmers. It shouldn't be anything else—ideally, not even prepared or processed foods. But practically, you can't do that. Because in the dead of winter, when there's only kale, chard, beets, and winter root vegetables, maybe apples that are coming out of cold storage for the third month, you've got to give people a reason to come to the market. So there is this balance. But where do you draw the line? We live in Northern California. If you go to the downtown Santa Cruz Farmers' Market in the dregs of winter, you can get actually pretty blown away by what you can buy in the wintertime. You get out of the state and the whole game changes. It's a struggle just to get people to come to the market, even during the height of the season.

I work about seven days a week during the high season, from April to November. I don't officially ever take a day off in the high season. It's just impossible. I'll take two, three hours off during the day to do something different. During the down season, I try to take a concentrated amount of time off, like two to three weeks, or if I can pull it off, a whole month. I'll go travel and just be completely void of the farmers' market stimulus so that then I can have the courage to do it for one more season. [laughs] It's really important. And farmers will say the same thing. At the end of the season, they're tired of the pain and suffering. [laughs]

I get up at five pretty much every morning. I've got a pattern. I drink my coffee; I read for awhile and then I hit the office probably around seven-thirty in the morning and I work all the way up until I can't work any more. Then maybe I'll take a couple of hours off during the day, and I'll come back and finish anything up. So my days will be anywhere from six to ten-hour workdays. It just depends on the time of year and what needs to be done.

The market is a membership-based nonprofit that operates on an annual cycle. There're markers throughout the year, like annual applications—and getting crop lists all set up. A lot of it is phone calls going in and out, contacting farmers about harvest dates, production problems, scheduling. Tons of e-mails. You get a flood of people wanting to get into the market. They want to sell anything from hula hoops to persimmons, and you've got to filter all of that. I'm the gatekeeper. So it's a lot of diplomacy on the phone, being firm, but also listening to people, because they want to sell a product. They'll do anything they can to convince you that their selling the best product on the planet, no matter what you're telling them.

When I came on, I worked with our Board of Directors and our attorney to develop a new commodity selection process based on the priorities of the market. We also use this system to select new vendors for the market association. The system allows us to set up a contract with the vendor.

Contracts will sometimes change three, four times throughout a year. But you as a vendor have a contract that you're legally liable for. That prevents the vendors in the market from selling something they're not approved for. So, for example, strawberries and tomatoes, are two crops that can be very profitable and everyone wants to sell them but we only allow a certain amount of growers to sell them at any of our markets to support a healthy supply. That can change depending on the demand for a product or if the supply changes. Without the contract-based system, we would not be able to regulate the item.

The biggest fraud problem is supplementation. A farm that typically grows one acre of tomatoes, but is actually selling two acres worth of the product at the market, is probably supplementing. Where did that other acre come from? Probably from a packing house. So that's one type of fraud scenario.

We've kicked out various people for cheating. Things like—this is the blatant, obvious one, and if you get caught with this, you just deserve to be caught—showing up with stickers on your fruit, which strongly suggests that the item was purchased, not grown by the seller. Or people will push seasons on the front end and the back end. Like, for example, asparagus historically comes in about March, maybe sometimes April. South County will get it first, and it starts to move up. What if you show up with asparagus in February? If you can get a two-week jump on competition for a hot commodity like asparagus, and you're buying and reselling it, you can sell it for a premium, and you didn't have to put any labor into it. If I start to notice that, then I'm going to start paying attention. I'll start making phone calls. I'll get the ag department involved. I'll call for spot inspections.

I've worked on the implementation of the new electronic benefits transfer system, which has improved the redemption rate of food stamp dollars in our markets and also brought more people into farmers' markets. There's infrastructural cost—you have to create banking systems, accounting systems, and you have to create scrip. You have to replace the food stamp dollar with something.

Back in the old days, food stamp dollars were universally accepted. Then the federal government decided to get rid of the paper scrip because of too much fraud and other problems, and they wanted to go to an electronic form. So they issued these credit cards called EBT cards. Each state's got one. These particular cards can be used at any point-of-sale machine, which are found in any store. But we set up in parking lots or streets or wherever a farmers' market sets up. We have no infrastructure. We have nothing to plug into. So that became a problem.

So the state said, "Okay, what we'll do for farmers' markets or other organizations that have this problem is if you do a certain amount of redemption per month, we'll give you a free machine." Fortunately, we redeemed enough food stamps under the old program that we qualified for a free machine. We still had to buy the scrip and get it printed. There are specifications that the government had for the scrip. We had to create our own accounting system. And we had to staff it. So you have to add labor costs.

We developed this system, and I have to say, it's a really good one. It's easy to run. Anybody can learn it. The great thing about it was, our redemption rate over the course of about two years went from seventy-five, to one hundred, to three hundred, to four hundred. Now we're hitting five, five-fifty.

If a farmers' market is set up to offer the shopper value, I think it's going to be successful. If it's real high-end across the board, then it's going to be subjected to economic cycles. Are farmers' markets more expensive than other retail produce sources in general, as many people believe? That's the million-dollar question. People are like, "Whoa, that's too expensive! I'm not going there." They're right and they're wrong. If you shop around in the market, you're going to find good deals. Yes, at times, it will be more expensive, but keep in mind that farmers' markets are subject to price fluctuations much more than the stores are. So I say, just go shop around. Be proactive. Don't go spend your money right away. That's how you do it in the Old County. Go shop with my aunt in north India, man. She walks around three or four times in the farmers' market. She haggles a deal, and then she comes back and pays for the best quality for the best price. Us as Americans, we're not used to that. We'll just go to the store and pay the price. The farmers' market offers a much more flexible opportunity. It's actually much more free-market.

The other answer is practical, but it has a philosophical element to it. You're going to save money on the front end, but you're actually going to be spending a lot more money on the back end. Cheap food typically comes at a hidden higher cost—cost to the environment, cost to local jobs, and possibly a cost to your health.

The paradigm by which the wholesale, centralized distribution food system works relies on cheap oil and supports a few large operations. That's putting a lot of faith in the hands of a few. By supporting a diverse local venue, you're actually creating more security for yourself, because you're supporting an infrastructure that is better equipped to deal with shocks and ultimately has a better chance to survive.

If you take away dollars from local farmers, then they will stop farming. If they stop farming, then you're not going to have local farms. If you don't have local farms, then you have to go to the outside area for food. That creates risk. And the more risk you have, then the more subject you are to systemic shocks.

Let's just look at the case of oil as a grand example. We have to go and start wars in other countries to secure oil fields, because there are systemic shocks that are coming down the pipe. We're tapping all of our oil from cheap sources, but it comes at the expense of our domestic security. Food is no different. As we move into this new paradigm of the global food system, food production is being controlled by fewer people and we're growing our food farther and farther away, well, what if the shipping lanes get blocked? What if there's a big storm? What if they just dump more pesticides than they should? There's a lot of inherent problems with supporting the old system. Whereas if you support a decentralized local system, you know what you're getting and who you are getting it from. You're looking the farmer in the face.

That sometimes falls on deaf ears, because it boils down to: "Well, it just costs me more, and I've got to feed my family." I don't know how to solve that problem right now. I think that WIC and EBT really does help, but it's not a solution. For those folks, the cost of food has to come down at the farmers' market. But then, at what expense? You live in the state of California. Do you know what land prices are? The cost of living, the cost of doing business in this state, it's fixed. I mean, if you lived in a different region, maybe it would be cheaper. It's hard to have those conversations with people.

We're looking for smaller scaled sustainable-based farming, not large-scale operations. We also look for growers that are certified organic. Certified organic is the only thing that's defined, and there's paperwork behind it. There's accountability. We also look for production-based farms, not backyard growers.

However, if we have an application from a farmer claiming to be using sustainable farming practices (not certified organic), versus another from a certified organic grower, we might go back to the sustainable farm and say, "Okay, you need to describe exactly what you're doing with your farming practices."

If it is determined that the non-certified organic farmer was a better choice, if we they felt like it was a more progressive approach to farming, whatever the reason, we would possibly choose that candidate and deny the certified organic grower. So there's flexibility built into the system. We did that on purpose, because we knew that this is just an unwritten book. But as a general rule of thumb, we're looking for people who are gravitating away from petrochemicals.

We also select growers that are closer to the farmers' market center, and move out from there. But we don't give it a 150-mile radius, because things like stone fruit or citruses might fall a bit out of that radius. In order to have a successful farmers' market business, you've got to have enough crop variety to make it work with customers. You have to be flexible, but the emphasis should always be on local production first. It would be ridiculous bringing in strawberries from Santa Maria, which I've seen in farmers' markets in the Monterey Bay. It's like, wait a second. The last time I checked, the number one agricultural product in the Monterey Bay area is strawberries! And you're going to Santa Maria?! If we're pulling people from who-knows-where, our customers are going to be like, "Wait a second." They understand. This is an educated population. They understand the value of supporting local food systems. So we'll continue to do that.

Getting into farmers' markets as a new farmer, that's a tough one. It seems that everybody who graduates from the apprenticeship program at the UC Santa Cruz Farm and Garden wants to stay here. But there's truckloads of other people that are already farming in the area. There's only so many farms that can be productive locally. There's a carrying capacity. You can only have so many farmers' markets in order to have healthy markets. So there's a carrying capacity to that too. And when you hit the max on both, then you have to reevaluate and say, "Well, maybe I need to be in a different region."

The older farmers know what's going on. They've been through it, and if they're still farming after twenty years, they're not calling me up saying, "How do I get into your market?" They understand you have to go out of the local area. You've got to sell at other farmers' markets.

But at the same time, start growing commodities that could be sold in your local market. Eventually, some of the older guys will retire, and there'll be opportunities for young people to come in.

I have a lot of young farmers come to me. They get frustrated, and they take offense. They take it so personally. "Hey man, don't take it personally. You're doing good work. It's just we don't need these products, and we're not going to bring you in and then create disturbances for other people. It's not going to benefit you. It's not going to benefit them. If this is something you want, then you have to be patient."

The way to broaden the carrying capacity is to increase the customer base of the farmers' markets. How do we get more people into the market? Everybody's got an answer: "Well, just get music." "Just do this." I'm not sure it's a simple answer, but I'm always trying to figure it out. The café seating at the Westside and Live Oak Market, well, that's one approach. Live music is another approach. Diversity of product. Great parking. Cool advertising. Everything you could possibly do to get customers to come in, get excited, and then hopefully, they set a pattern, "I'm going to support this." But I think the pivotal way for them to start the pattern is that they intrinsically believe in what they're doing.

People are pattern-oriented. You've got to get them interested in what they see, and then you've got to get them to develop a pattern. Coming to the market a few times because it sounds nice and it's cute is not really what the farmers' market needs. They need core customers that are spending twenty, thirty, forty dollars per week, fifty, a hundred. We need to create core customers that the farmers can rely on. And that is just going to take time and education.

Guys like Michael Pollan have done wonders for the farmers' market business. One of his books does more than five years of my advertising. It's getting the word out. It's programs like the UC Santa Cruz Farm and Garden, Community Studies, agroecology. It's educating the youth about the importance of local food systems.

The local high school systems need to get on board. There needs to be more of a connection with the farming community. Young people need to have an opportunity to get their hands in the dirt, and at the same time, to try to run a farmers'-market business. So I have interns all the time. And I give them a crash course in the reality of what we do for a living. Then they go out and tell other people. That's just how it starts.

You go to a retail outlet with fluorescent lights, you don't feel like you're participating. It's like, I just went to the gas station, and you just check it off your list. I got my oil changed and went to the store. There's no heart to it. When you go to Windmill Farms' stand at the Downtown Market, Ronald Donkervoort is covered in dirt, because he's been picking beets out of the ground that morning. He was out in the fields. Or the strawberries that people just die for, he just harvested right before the farmers' market. I mean, I know his operation. Where are you going to replicate that? Nowhere! And that's where a farmers' market has got everybody beat. I promote that freshness.

And you're talking to the farmer. You're meeting the person who grew your food. You can't do that in the organic section of a store where you're buying Earthbound [Farms] lettuce. But at the same time, I'm glad those people who don't have access to farmers' markets have access at least to foods that aren't sprayed with who-knows-what.

The core mission of the market is to support small-scale, local, sustainably farmed agriculture. That's it in essence. We create a platform by which farmers can come and meet in a safe environment and run their business and be totally supported. These are the guys that are doing the hard work. The farmers have got to be the focus. If we support them, then we're doing the right thing.

What I love most about this work is seeing the whole concert come together. You go to a Wednesday market in September and all the cylinders are firing. Everything is on the table. There're gazillions of people buying their food. The farmers are doing their thing. It's this whole energy that just takes over, and you look at it and go, "Oh, my God. This is so important. This interaction is so fundamental." It's not buying a new car. It's people buying food. They have to have it, and they're not willing to grow it themselves, so this is the next best step. You see it all come together under one roof. And you're just like, "Whoa, this is amazing." That's some of my happiest moments, definitely.

I think the Downtown Farmers' Market in Santa Cruz needs to have a permanent infrastructure developed in concert with local developers, the city of Santa Cruz, and the community at large. There's chitchat in the city about moving us or building a parking structure where we're at. My primary fear is the city not focusing on the essence of what we do. I've been working on this for a couple of years with these guys. I see other towns where the farmers' market is so embraced.

The city is bending over backwards. It seems kind of ironic that the capital of organic farming, Santa Cruz, isn't more willing to have discussions with us about these things. But we'll figure it out. It will come together. I think we're moving into a golden era for farmers' markets. I want to keep the bar raised really high so people will trust and respect the institution and what we represent.

Larry Jacobs

Photo: Tana Butler

As a young man, Jacobs owned and managed a tree nursery in Southern California. When aphids infested some of his trees, a pesticide inspector sold him Metasystox to apply with a backpack sprayer. Jacobs temporarily became very ill from pesticide exposure. Vowing never to apply pesticides again, he searched for alternatives and found a mentor in Everett ("Deke") Dietrick, a pioneer in integrated pest management.[1] Jacobs left the nursery business to study soil science at California Polytechnic University, San Luis Obispo. After graduation, he moved to Maine to apprentice with Helen and Scott Nearing, grandparents of the back-to-the-land movement. In Maine, Jacobs met his future wife, Sandra Belin. They eventually moved to the town of Pescadero, California, where in 1980 they founded Jacobs Farm, now the largest organic fresh culinary herb producer in the United States. In 1986, Larry and Sandra began working with a group of family farmers in Mexico to start the Del Cabo organic growers association. Together they created an international market for organic vegetables grown in Baja California and Jacobs Farm became Jacobs Farm/Del Cabo. In 2008, Jacobs won a landmark pesticide drift case against pesticide application company Western Farm Service, Inc., now Crop Protection Services.[2]

---

[1] For interviews with "Deke" see http://www.rinconvitova.com/dietrick_interviews.htm

[2] On September 25, 2008, a Santa Cruz Superior Court jury awarded a one-million dollar settlement to Jacobs Farm. See "Santa Cruz Organic Farm Wins Pesticide Suit," *San Francisco Chronicle*, September 30, 2008. The court found that the contamination of non-target crops caused by the evaporation and movement of pesticides after application is the responsibility of the applicator. See also, "Appellate Court: Santa Cruz Organic Dill Grower Has Right to Sue Neighboring Farm for 'Pesticide Drift,'" Kurtis Alexander, *San Jose Mercury News*, December 23, 2010. http://www.mercurynews.com/ central-coast/ci_16923749?nclick_check=1. Jacobs won the case on appeal in December 2010. The appellate court decision is available at http://www.courtinfo.ca.gov/opinions/documents/H033718.PDF

I was taken by the soils and the town of Pescadero, California. A coastal valley struck me as the right place. In 1980, we began growing crops appropriate to that climate and time. We planted fava beans and learned to dry-farm peas. We experimented with more intensive crops like broccoli. We grew summer squashes and lettuces. We began marketing in the South San Francisco Produce Market and selling gourmet vegetables. Some of these crops became popular, like the baby spring lettuce mixes.

We had an eye for niche markets. Sibella Kraus at Greenleaf Produce encouraged us to grow culinary herbs. We purchased one hundred tarragon plants from a San Diego nursery. Water constraints made culinary herbs a good choice because they require less water. But the market was limited. People don't use very much rosemary, oregano, or thyme. And in 1981 or '82 and '83, there wasn't the interest in cooking with culinary herbs there is today.

We started small and gradually grew. We supplied restaurant distributors. We learned another advantage to culinary herbs, we didn't need a big truck. A dozen bunches of thyme doesn't take up nearly as much space as a dozen beets, though a similar amount of time is needed to harvest and pack and they had similar value. [Today] Jacobs Farm plants 240 acres in two California counties. It started as a small family operation and still maintains the family aspect, but the family has grown.

In 1984, Helen Nearing visited.[3] She was en route to a self-awareness course in Barra de Navidad, Mexico, and invited us to come along. It was wintertime. We were pretty much done for the season and offered to drive, as I had always wanted to see the Baja California peninsula.

Sandra and I left Barra de Navidad after a few days and took the ferry from Mazatlan to Cabo San Lucas and drove on to San José del Cabo. Instead of going to beaches, we walked and visited farms. We discovered a patchwork quilt of small farms, and good farmers, with the classic problem of a small, isolated market. When one person had cilantro, they all had cilantro. When one person had mangos, they all had mangos.

Origind del Cabo

---

[3] See the website for The Good Life Center at Forest Farm, which carries on the Nearings' legacy: http://www.goodlife.org/wordpress

They would all grow the same things at the same time. The sign going into San José del Cabo said the population was ten thousand. Farmers and extended families were part of that number, so who's going to buy what you grow?

The land was desert on the Sea of Cortez. A mountain range runs up the middle of the peninsula. The farms were in the large San José drainage. They had wells developed by the government. Concrete canals distributed the water. The old-timers told stories how they used split bamboo to move water from artesian wells down the arroyo to irrigate their farms.

As Sandra and I drove through the desert, our previous experience doing community development projects in the western highlands of Guatemala and working with small scale farmers resonated with what we had just seen. When we saw the sign for "Pescadero" an hour and a half out of San José del Cabo, we thought, "Pescadero North, Pescadero South!" That idea dominated our conversation all the way back to Northern California.

It didn't take us long to return to San José del Cabo. We stayed at the Hotel Ceci across the street from the church and plaza. In that little hotel room, we wrote a one page project proposal, which we presented to the director of rural development who we found in a small, dingy office in the municipal building. We proposed to create economic opportunity by teaching organic farming practices to local growers and linking them to winter markets to fill a void in the organic marketplace in California.

produce for -inter market in US

The rural development director was Narciso Agundez. He is governor today. He introduced us to the president of the ejido, a land management organization common in Mexico.[4] Years later, we learned that the ten individuals at our first meeting thought that it

---

[4] Jacobs added the following written comment: Ejidos were established after the Mexican revolution to return land to peasant farmers, campesinos. Land expropriated from large haciendas and of low value was given to ejidos. The ejidos were composed of small-scale farmers entrusted to manage and use ejido lands. Some ejidos operated communally. Others divided land into small parcels and drew names out of the hat to assign parcels. The land was yours as long as you farmed it. When you stopped farming, if your kids didn't want to farm, it returned to the common pool for somebody else. In many ways, it was a system before its time because it preserved farmland. You couldn't buy it or sell it. But you could farm it. The San Jose ejido had set aside a large parcel of land for communal pasture, as well as creating small parcels for individuals to farm.

was absolutely crazy to try to grow crops without pesticides, that I had *no concept* of how challenging and difficult pest management would be and how many different insects inhabited their tropical desert.

What they didn't know was that we were to succeed that first year. We grew squashes, green onions, tomatoes, a little bit of basil, some cilantro, and bell peppers. Everything did okay. But we managed to pull off the squash crop really well. Prior to planting, we developed a relationship with two Bay Area distributors, Veritable Vegetable in San Francisco and Santa Cruz Trucking in Santa Cruz. Sandra and I visited Veritable Vegetable before we returned to Baja and talked to Bu [Nygrens] and Mary Jane [Evans]: "We've got this nutty idea. What do you think? And by the way, if you think it's a good idea, tell us how many boxes of green onions, tomatoes and zucchini you would buy per week."

They were incredibly helpful and spent a lot of time helping us develop a plan: "Here's what you can grow." That's a grower's dream but we really didn't have it thought out that well. How were we going to get it all up here? We certainly weren't going to put it in our pickup and drive it. They weren't going to send their truck to pick it up. But if it wasn't for Veritable Vegetable and Santa Cruz Trucking it wouldn't have happened. They gave us a lot of support, encouragement, and guidance as to what and how much to grow. But what nobody knew, not the growers, buyers, or us, was that our first crop would be wildly successful, and the farmers would make more money than they ever had before.

Cabo San Lucas had a fishing camp, a basketball court, a little place to buy cold beer, a fish cannery and the ferry terminal. That was it. They had never seen sunburst squash, and they'd never seen the kind of cherry tomatoes that we were having them grow. Basil? They thought of basil as a stomach tonic, a medicinal.

These squashes were susceptible to viruses, and plant viruses are vectored by piercing, sucking insects like aphids. Aphids were on every plant! We concocted a mixture of sugar, dried seaweed, and milk that we sprayed on squash fields. I was trying to attract beneficial insects and had read that milk might inhibit virus transmission. We needed a sweet protein solution that nectar-feeding wasps and other beneficials would come to. We were flying by the seat of our pants. The growers said, "Oh, here comes Larry and Sandra with their coffee and milk concoction." *Café con leche*. I'd say, "Now, don't drink it!" We drove from field to field with fifty pound sacks of powdered milk, sugar, and dried seaweed. We mixed it in the field and sprayed it on.

We didn't have enough money to buy packing boxes. Fortunately, the ejido had a pile of pre-cut wooden slates leftover for mangos, but the mango box was too big. We needed small boxes to pack squashes and tomatoes. I asked, "What about that abandoned pile of wood?" "But they're too big." "C'mon!" All the guys got together with saws, hammers, and nails, and we're hammering together little boxes.

The next big problem was transportation to California. "Oh, there's airplanes." The airline staff in the downtown Cabo office were encouraging and excited about new cargo business. On February 1, 1987, we drove eighty seven cobbled together wooden boxes of squashes and tomatoes to Mexicana Airlines. And Mexicana Airlines loaded that first shipment on their Boeing 727 jetliner and flew off to San Francisco. The product arrived strewn all over the belly of the plane. The boxes vibrated apart. The cargo night crew gathered together the squashes and tomatoes from the plane floor. They lent Sandra a hammer. She began nailing boxes together. And the guys from Mexicana Airlines helped. They repacked the squashes and tomatoes back in the newly nailed together boxes, and Veritable Vegetable picked up the first organically grown summer vegetables produced in winter.

It *was* a miracle we survived that first season. There was a lot of help from a lot of people. The beneficial insects must have liked our milk and sugar concoction. The San Francisco-based Mexicana Airlines staff, all Latin Americans, heard about the benefit to families in Cabo and wanted to help.

Nobody in Cabo knew how to export. We were directed to the head of Customs, who we found in an old downtown office of plastered adobe walls. A large skeleton key opened the heavy wooden doors hung with rusty metal latches. The customs director, Don Arturo Gastelón, was an elderly gentleman. I sat down with him and explained, "I'm working with these farmers, and we got a problem. We're ready to harvest, and we want to export." He seemed to be listening but he didn't say anything. And I thought maybe he didn't hear, so I repeated myself a little louder. His face brightened, both arms shot up in the air, and he shouts, "Great! This is great!" And he told me how they used to grow tomatoes in the thirties. They would haul them on horses and mules from the farms down to the beach, and then row them out on boats to ships anchored in the bay. They packed the tomatoes in town in wooden boxes three layers high, picked green with a white star on their bottom, 'Ace' tomatoes, wrapped individually in tissue paper. Some townspeople still had the wrapping paper and ironically, the ships loaded with tomatoes were destined for San Francisco.

Don Arturo was really excited about the prospect of tomatoes once again being exported. I asked, "Okay, what do we have to do? How much does it cost?" He said, "Just come back when you're ready. Don't worry about a thing."

I returned early the next morning. The plane departure was at noon. I had a list of what we had packed. Don Arturo asked, "How much does it weigh? What's the name of each product?" So I wrote it all out. Then he pulled out five pieces of paper and four pieces of carbon paper, plopped down the typewriter, rolled the five pieces of paper and the four pieces of carbon paper into the typewriter and proceeded to create a provisional export form complete with borders and blocks for data. He used the small letter "m" to outline each block. It was artistic but took more than two hours as he told me stories about the past. And now it's getting late. I had to get to the cargo office to document the shipment, and then to the airport. I thought I had plenty of time but hadn't anticipated how long it would take with Don Arturo.

It was getting to be eleven o'clock. I asked, "Arturo, when are we going to get done?" He said, "We're almost done." Finally he hands me the "provisional" export documents, *pedimiento provisional,* after he stamps it three or four times and signs the bottom. I ask, "How much?" He says, "Nothin.' Get goin.'" I grabbed the papers, and dashed down the street to Mexicana's cargo office. They asked, "Where have you been? The plane leaves in an hour. We can't hold the plane for this." They documented the shipment and filled out the air waybill, and we all raced to the airport. We got to the airport in record time and loaded the boxes on a baggage trailer. We rode with our cargo of squashes and tomatoes on the back of the trailer to the side of the plane, where we helped load the plane's cargo and luggage compartments. That was the first shipment. Today we're not allowed onto the tarmac, let alone to load the plane.

I went back to Pescadero in June. It was time to plant in California. A few months later Sandra and I put together packets of seeds, sent them to Cabo, and said, "Look, you guys know how to do this. We've taught you how to grow these crops. We showed you the spacing, at what stage to harvest, and how to pack. You now know about beneficial insects." We had compost-making classes and discussed the importance of organic matter and how to recycle. The county was providing slaughterhouse waste, and we were working on getting the green waste from town. We had done soil tests. We were growing corn and black-eyed beans as cover crops and worked out how to lower the soil pH to get better crop growth. It felt like a solid foundation for going forward. We'll just send them the seeds and off they go. We sent the seeds, and we got back this call, "No, you gotta come back!"

We had a farm to run in California. We thought we'd seed the idea and be through. We were going to show them how to get organized. We'd show them how to divide up the receipts. We set it up to be fair and equitable, and we thought it would fly. And they said, "No, you gotta come back."

So we returned, and it became a full-time endeavor. For many of those beginning years, Sandra stayed in California running Jacobs Farm, interfaced with customers, and managed the importation.

I learned about sandy soils and different kinds of irrigation in a desert at the Tropic of Cancer. It was a real change from the alluvial clay soils and the Mediterranean climate we were working with in California.

Then there was the question about organics. We wrote a one-paragraph declaration, "I, Farmer José, swear on a stack of hoes that I'm farming organically and growing plants without synthetic chemicals. Here's the date. Here's my signature, and here is the witness." My wife or I would sign as witness. Every year, we'd walk every farm and everybody signed an affidavit. It was "official." But what really worked was social pressure. It was understood that if you broke the rules, you ruined it for *everybody* because the companies who we were selling to, the people who were eating what we grew, were buying it because they wanted food free of toxic chemicals. If you broke that trust, you would destroy your business. So there was enormous social pressure to maintain the group's integrity. They began to understand the importance of organic food for their own health and the environment. Some farmers would explain to visitors, unsolicited, why growing organic was important. They owned the idea of organic farming and its importance, and they were teaching other people.

We started in 1985 and '86 with ten families. So we've been doing this for over twenty-five years. Today there are between four and five hundred families spread out amongst eight different farm groups up and down the peninsula from the tip of the Baja peninsula in the south to Ensenada, sixty five miles from San Diego, California. We've tried to select communities with economic needs and who didn't have the economic wherewithal to do something on their own.

Today much of what is produced is trucked, though we still fly a lot of basil. You can no longer ride with the product to the plane and load it yourself. Since 9/11 new security on both sides of the border has made everything more bureaucratic.

We're now planting two to three thousand hectares. That produces a lot of organic food which feeds a lot of people.

Besides the four to five hundred families directly involved in the growing, there are many more individuals involved in packing, shipping, logistics, and administration. There are accountants, agronomists, plant breeders and entomologists. It's a business that benefits many people. Sandra and I have become cheerleaders, advisors, and orchestra conductors. Through Oregon Tilth, we have reciprocating certification for Europe. Products are occasionally flown to England and the Middle East.

When we started shipping by plane, we were asked about the environmental impact for moving produce such long distances out of season. We saw it like hitchhiking and taking advantage of unused plane space. The planes were going there anyway and returning empty. We tried to pick and ship the same day. It arrived in San Francisco the same night and was distributed the next morning pre-dawn. It could be in the store the next day. That was pretty darn fresh. Planes continue to fly tourists who are escaping winter's cold and rains. Tourists aren't taking lots of big bags. Airplanes had quite a bit more carrying space than just the tourists and their swimming suits, so we were making good use of what was already going and coming. We were carpooling. But instead of carpooling people, we're carpooling boxes of basil and cherry tomatoes.

Trucks are similar. The Baja Peninsula has become a magnet for tourism. It was targeted by the Mexican government to be developed as a tourist mecca and so it has. To supply the hotels and restaurants, trucks go south with supplies. Those trucks look for loads going back.

Some things you can't buy local, and you've got to make a choice. Do you give up your coffee and chocolate? What about local bananas? The crops that we're growing in the wintertime, you can't get local in New York or Boston when there's three feet of snow outside, or in the Bay Area during winter. They just don't grow. If we choose to be seasonal, that will be fine. The Cabo farmers who have grown up with Del Cabo will be fine because there is now a growing local market for their crops and a burgeoning farmers' market.

Twenty years ago we were not recognized by the agricultural community. The local San Mateo County adviser told me, "Larry, you are nuts." A neighborhood grower said, "In a few years you'll be spraying chemicals too." Upon meeting me, Noel Diaz, who was in his nineties or late eighties and farmed in Pescadero, said, "Yes, I know who you are. You're the young guy who's organic. What's wrong with you, anyway? There's nothing wrong with pesticides. I can eat DDT all day long, and it doesn't hurt me." A few years before Noel died he asked, "How do I get my farm certified organic? I stopped using pesticides." That's how much it's changed.

Those individuals who chose to farm organically before it was fashionable, when it was considered fringe and kooky, and are still farming, are doing well. Over the last twenty years there's been a growing public awareness about the benefits of organic food. In 1989, we had the Alar scare. Since then organic food and the impact of pesticides has been regularly reported. The public is better informed. And organic farming is a recognized and important part of our food system. It was tough getting here.

I'm thankful for all the organic farmers that stuck with it and made it happen. We wouldn't have organic food if there hadn't been a committed and dedicated group of individuals who said, "We're going to grow organic food because there're people who want it."

Dick Peixoto

Photo: Ceasar Garcia

Dick Peixoto (pronounced Peh-SHOTE) exemplifies a recent type of organic farmer who, after a long career in conventional farming, transitions to organics for a mixture of reasons. Attracted by the organic price premium, Peixoto decided to transition to organic farming and began Lakeside Gardens on a fifty-five-acre farm in Watsonville in 1996. His conventional farming friends thought he had "lost his marbles," but Lakeside Gardens is very successful and Dick has become a spokesperson for integrated pest management, hedgerows,[1] and other organic farming methods. The company has expanded its operation to a total of twelve hundred acres, including fifty different parcels in the Pajaro Valley, many of which border on hospitals and schools trying to reduce pesticide exposure. Lakeside also farms on five hundred acres in El Centro, making it one of the larger organic growers on the Central Coast and in California. All of the company's produce is California-grown and shipped by Albert's Organics and other organic food distribution companies to grocery stores such as Safeway and Kroger's, as well as Whole Foods.

I was born right here in 1956, and grew up right here in Watsonville, California and never left. My grandfather on my father's side moved here from the Azores Islands and farmed potatoes here in the valley back in the thirties. My father did not do farming. He got into selling fertilizers, pesticides. I started the farming again on my own.

---

[1] See Sam Earnshaw, "Hedgerows for California Agriculture: A Resource Guide," (Davis: CA, Community Alliance with Family Farmers, 2004). www.caff.org/programs/farmscaping/Hedgerow.pdf

We had a fifteen-acre ranch out on Green Valley Road. We moved there when I was ten months old, moved out of town. There were seven of us kids, and we grew up on the ranch. We had apples and cows and chickens, not so much for fun but more to save money. We slaughtered our own beef and milked our own cows. When other kids, city folks, would get up to go to school at seven or eight o'clock, we had to get up at six o'clock, go milk the cows first, then go to school. At the time, we thought we were being cheated, but looking back, I think it was a great experience for us to grow up having chores—everything from milking cows to shoveling out the corrals. I used to pick apples for my dad. He used to pay us five cents a box to pick apples. We learned a lot about work ethics, all us kids. Not one of us has ever been out of a job, because we learned the value of a job and worked all our life.

Probably the biggest influence in the direction I took in my life was driving with my dad when I was seven, eight, nine years old as he worked for a fertilizer/pesticide company here in town. We rode around selling fertilizer and insecticides. We learned a lot about chemicals and sprays—the good and the bad, and everything in between. We grew some raspberries, and we grew apples. As kids, one of our jobs was pulling spray hoses. Back in those days, we didn't have the tractor to spray, so a guy would come in with a big truck with a 500-gallon tank on it. He would charge my dad extra if he had to drag the hoses, so us kids would go out and drag the hoses for him while he sprayed. At the time, we weren't thinking about being poisoned or anything. We were just thinking it was a job. So we spent our times—you know, twelve, thirteen years old—dragging hoses for Jim Dalton so he could go spray the apple trees, and that was part of our job. We learned a lot about how they spray, and how to prune, and how to thin—a lot of the jobs that revolve around the farm.

[There] didn't seem to be [safety concerns about pesticides] back then. When I was probably thirteen, fourteen, I used to pick blackberries for one of the local farmers. At lunchtime we had to break early because the plane was going to come in and spray the field. After lunch we could go right back to work. So it wasn't even thought to not go back out there after it was sprayed. We'd go back out and find dead rabbits and birds out in the field. At that time, the pesticides were ten times as bad as they are today, but we thought nothing of it, and nobody really had concerns about it. I don't remember being sick from it. I remember seeing those dead rabbits and dead birds, and I thought that was kind of interesting.

We started Lakeside Organic Gardens in '96. My big motivation [in transitioning to organic] was survival. There're different kinds of sustainable. I could see that the operation would not sustain itself because of all of the pieces of the pie being cut off of the business. At that time, we were one of the sizeable head lettuce growers, probably in the top three in this valley. We did about 2,500 acres of head lettuce conventionally, and had been doing that for years, but still, our margins weren't that good. In each year the pack charge would go up, the carton cost would go up, the sales commission would go up. We grew the crop, and then all these pieces of the pie would be taken off the plate. If there was anything left, we would get it. But sometimes there wasn't anything left.

So back in the early nineties, I started watching these smaller organic growers. Frankly, I thought their farming was terrible. A lot of them were very small growers, didn't have the proper tools. I said that if we took the tools that we have today and the operation we have today and applied it to organics, I think we could really make things work. It wasn't quite that easy. That learning curve was real steep. I thought the big thing was going to be handling the insects, handling bugs, but it really wasn't. That was something we learned fairly quickly. The big problem for us was conventional weed control and soil fertility. Our tools are limited. In the old days, with weed control you could go out and either fumigate with methyl bromide and annihilate every molecule of everything that's out there and start with clean ground. Or they had these herbicides that you spray and eliminate ninety, ninety-five percent of the weeds, so you really didn't have a big weed problem. Organically, all you have is mechanical tools.

We've developed things over the years. Like we have a propane burner that burns off the weeds and different tools like that that we're able to use. We have a tractor, and we have hand burners. Actually, like, in the wintertime, if it's too wet for the tractor, it's a lot cheaper to walk through with burners and burn weeds than it is to later come back by hand and pull each weed by hand. Conventionally, you can swing a long-handled hoe all day long because there're no weeds to deal with, but organically you have to get on your hands and knees and pull every weed by hand. There's no other option. A lot of times our cost on thinning lettuce—conventionally it would be eighty, ninety to a hundred dollars per acre. Sometimes organically we'll be up to five to six hundred dollars per acre because of the weeds. So it's a big difference. It's not a little thing. It's a huge thing.

It took us a while [to transition to organic], [because] a lot of the crops we grow are very tough to grow. Like, celery is one of the ones that I thought we'd never be able to grow organically because normally we'd spray between ten to fifteen times to grow a crop of celery. We'd spray for an insect called leaf miner. I got really frustrated at the amount of dollars that we spent to spray our crop to bring it to harvest. Now that we've gotten into organic, with the beneficial [insect] program, the beneficials come in and fight the battle for us. We've never sprayed a block of celery for leaf miner, and we've never had a leaf miner problem. The other thing, too, is celery has a high requirement of nitrogen to grow the crop, and we would normally just pour all the fertilizer we could to it. But we learned how to grow celery now with probably half the fertilizer that we were doing conventionally, and as of this last year, we've actually surpassed what we did conventionally, production-wise. We're producing more units per acre organically than we were conventionally. I never thought we'd do that. It's taken us a while to learn how to do that, but now I'd say the quality of our celery is as good as conventional. People say our celery tastes a lot better than the conventional. The conventional has kind of a watery taste, and this has a real sweet taste to it.

About fifteen to twenty percent of our business [is direct to stores in the Bay Area and Central Coast]. The rest is all wholesale, which goes to the biggest customers, Whole Foods. And then we do Albert's Organics and distributors like Four Seasons Produce. And there're quite a few different chains. We go to Safeway, and occasionally Albertson's and Raley's, Kroger's, quite a few different stores. They all seem to be getting into organic. There're a lot of new people on line now that really want to get into organic. Everything changed about a year and a half ago when Wal-Mart decided to get into organic. Basically every chain in the country said, "Well, if they're getting into it, we have to get into it to compete with them. They're not just going to get into it for the fun of it." So I think basically every chain out there now is wanting to get into organic.

The thing is, is the market in organic is a whole different world to live in. The chain stores are getting an education, because before, they could just go out and order whatever they wanted, fill the shelf space up, get the cheapest price and put them on the shelf, and price it however they have to. It worked. But organically, you can't always get every item every day. There're items sometimes that there's none available.

We were just looking at Napa Cabbage this morning, and my brother says, "Well, geez, this is pretty ugly Napa Cabbage," and I says, "You know, Bill, that's probably the finest Napa Cabbage in the United States today. It might be the worst, too. Come to think of it, I think it's the only." So I said, "But we're probably the only company. We have not found anybody else in the whole country that has Napa Cabbage. So our Napa Cabbage is a little bit smaller and a little bit more yellow than we like to see it, but it probably is the finest Napa Cabbage in the country." And we're not afraid to charge for it because we lost the last three blocks of Napa Cabbage because of the cold weather. So we're going to try to make up for lost time. Usually most of the natural food chains, they understand the pricing goes crazy sometimes. It's not unusual for us to be three to five times the price of conventional.

Safeway made a stab at organics about four years ago and it was a disaster. Then they hired some organic people to run it, and now it seems to be taking hold, and they're doing this whole movement now for organic. They're advertising organic. They're pushing it. I'd normally think of organic as produce, but there's a whole other world out there. [Processed organic foods are] a huge market. We sell to some processors like Amy's Kitchen, and if you go into a lot of these stores, I'm shocked about how much shelf space Amy's Kitchen has. They're a little local California company but committed to organic. They don't do any conventional. I just saw an order today in from Amy's Kitchen. They're looking for chard because they have to have a steady supply, and with the freeze they have to shop around. There're not that many suppliers that can supply them with the amount of green chard that they need, so they called us to try to fill that order.

But, like I say, it's a very limited amount of people. If you wanted broccoli in the state of California today, you could probably call over a hundred conventional people. If you want organic broccoli, you have about five phone calls to make. Don't waste your time calling anybody else because all they would be doing is buying from those five people. So it's a whole different world. That's why I think Wal-Marts are really having a really frustrating time right now trying to deal with the whole organic field. Because before, they would just pound their fist on the table and say, "We're Wal-Mart and we want the end product." They would have people stand in line that could supply it.

[When I transitioned to organic] the conventional farmers thought I'd lost all my marbles. They thought I was nuts. They thought, really, I'd gone off the deep end. They thought I was just flat-out crazy. And a lot of the organic growers that had been telling me for years I should just stop spraying and get off the deal, they were kind of ticked off at me because when we transitioned, the ground we were transitioning was a hundred acres at a time. They said, "Dick, you're gonna run me outta business." I said, "Well, you guys were the guys telling us to go ahead and get into organics." So I was getting it from both sides at that point. But there's been a big swing, I'd say, in the last five years, about the average grower's mentality toward organics. I think most of them have really grown to respect it as a viable business. And frankly, to be honest with you, in the old days I saw guys farming organically, and they would go wade through the weeds, getting ready for the farmers' market on Saturday. They would wade through the weeds, and then they'd reach down, they come up, "Hey, I found a beet!" You know, it was just like this treasure trip to go through and try to find things to harvest. Whatever they could find in that weed patch, that's what you'd end up with at the farmers' market. And they got that reputation; if you want to farm organically, you just grow a lot of weeds and just deal with it.

If we were out standing in one of our organic fields right now, I'd challenge anybody to tell me if it's organic or conventional, because we found that we need to control our weeds, not so much for this crop but for the future crops, because every weed we allow to go to seed could be a thousand weeds next time. Like, I had a sizeable conventional guy the other day. They were buying organic product from us. He said, "I want to go out and look at the field," so we took him out to look at the field, and he goes, "You're lying. This is not organic. This just looks too good. There's just no way. It looks better than the conventional field I just left. There's just no way." So I showed him the certification and everything. I said, "Well, look at any field you want to. Here's the planting schedule. Go look at any field you want to." They were shocked. But it didn't happen overnight. It's taken us a long time to get to that point to where we can manage the pests, and manage the fertility, and manage the weeds.

The important skill is people management. If you get the right team of people doing the job, then they learn a lot along the way with you, so you don't have to be there for every step of the way. We have a production meeting every morning at six o'clock for about an hour. I meet up with the top guys—the guy that runs the tractors, the guy that runs the irrigation, the guy that runs harvesting. We meet every morning to try to work as a team to get the job done. I tell people on any given day we have five hundred blocks of produce growing out there. Each one has to be managed individually. So it's not uncommon for us to sit down at a production meeting in the morning and talk about six beds of radishes, or eight beds of radishes. Which bed of that field is the weedy one, and why is it weedy, and how do we do better? And did you get the rows wet? Can we go plant six beds of cilantro tomorrow? Does the Mei Qing Choy have good germination or bad germination? So we go through all of these things to try to figure out how to do a better job at it as a team effort. If you can manage that team, and they can manage the people that work under them, then you got it made [and] all the other things organically that we've talked about will fall into place, because you've got the team that's watching out for you.

I speak pretty fluent Spanish. I only went through high school education-wise, and in high school I took a couple of years of Spanish and goofed off and really didn't pay attention to it. As soon as I got out and started to be in farming, I was kind of shocked. I said, I should have listened because I need to know this stuff. That was kind of bad in a way, and it was kind of good in a way, because the Spanish I learned in class is a little bit different than the farm workers so once we get out in the field, we learn a whole other language. You have better communication skills by learning their language as opposed to book language, which they don't really even understand either, so it's [chuckles] kind of funny. The other thing, too, it's very important to communicate with the Spanish speakers in Spanish. I know of some growers that have had a real tough time in business because they don't communicate with their people, and so every time they want to say something, they got to go through a translator. The people don't have any respect for that. You lose the whole momentum along the way. Whereas if I go out and talk to a guy, I can talk to him about how's his family, his kids—are they playing soccer? A lot of times I'll give a guy a ride from one tract to the next tract. We can talk about when he's going back to Mexico, how's his ranch in Mexico. And that's important, for them to know that we have enough interest to take the time to talk to them about that. I've had some people work for me fifteen, twenty years. We don't pay the highest wages, but we pay fair wages, and we've helped a lot of these guys along the way buy houses or whatever. A lot of times they need help or temporary help, whatever. We chip in to help them out.

We've had major [labor] shortages. In fact, we were on the national news last year because Associated Press heard about our labor shortage and came down to do an article on one of our fields. I think that it's a compound effect. We're already starting to experience labor shortages this year because the strawberries have started up in the last couple of weeks. A lot of the workers would rather go pick strawberries, where they work eight, nine hours a day, and they kind of do the same thing, day in, day out. As opposed to our guys, a lot of times, because we're so diversified, the guy will go pick radishes for three hours; and then he'll go pick cilantro for two hours; and then he'll go pick dandelions; and then he has to go to the other end to pick parsley; and then come back over here and pick eight boxes of spinach or whatever it is. All of those, he gets paid piece rate, and he makes probably as good or more money than the strawberry guy, but it's so much easier in strawberries. You just push a cart down there and zzzhhht—you know? You're one of one hundred or two hundred berry pickers out there, and it's just an easier lifestyle. Raspberries, same thing, there too. Standing up, where you don't have to stoop over all day, just picking raspberries and make your money that way. And then there's times, like I said, with the weeds, where we have to organically, we have to get on our knees and pull every weed by hand. Nobody wants to do that.

So the organic people, I think, are affected more than the conventional, from that standpoint of the style of labor needs. All of our piece rates are higher than the conventional guys' because of the volume that we do, and sometimes there's weeds to deal with and things like that. So we adjust all the piece rates that we pay on all the harvesting based off of conventional plus extra because it's organic.

We have the cooler climate, so we're able to grow crops that can't grow anywhere else in the country. You can't grow cauliflower in Nebraska in the middle of summer, but we grow the most beautiful cauliflower here. We grow broccoli, Brussels sprouts, Bok Choy now, Napa [Cabbage]. A lot of the items we grow here you can't grow anywhere else in the country. And I think we get a lot more support from the community as a whole for farming sustainably in this area than we [would] if we were out in the middle of the San Joaquin Valley or something like that. Everything over there is focused on bigger farmers. Like I said, I've seen a big evolution in the last five years. A lot of people that were anti-organic now have switched around all the way to pro-organic because of what they see we can do. I think there was this kind of fear factor before. Are we going to have weed-infested fields, and are all the bugs going to come from your place over onto my place? Things like that. We just don't see that anymore.

I have three kids. I'll be honest with you. I really don't want to get them into farming. I love farming. I've always loved farming. It's a way of life for me. But you really have to have that mentality to put up with all the problems today. We're under constant siege from the regulations. There's a shortage of water; they're regulating water runoff. Then the air pollution board is after you for dust from the land that you farm. The planning commission is after you because your fence is too tall. And the Fish and Wildlife is after you because you disked the corridor along the edge, made a road through your ranch that happened to hit some tules. All of these regulations, the compound effect—it's a nightmare to try to outrun all those bullets. There's easier ways to make a living, I can tell you.

Dee Harley

Photo: Paolo Vescia

In the village of Pescadero, forty-five minutes' drive north of Santa Cruz, Dee Harley runs San Mateo County's only active dairy. Harley and her staff care for a herd of more than 200 American Alpine goats, crafting the animals' milk into sought-after cheeses (chevre, feta, ricotta, and fromage blanc) that have consistently won awards at national and international competitions. An increasingly popular agritourism destination for residents of the San Francisco and Monterey Bay Areas, Harley Farms also offers leisurely, informative tours of its entire dairy operation, from the birth of hundreds of kids each spring to the on-site sale of delicate white cheeses decorated with fresh herbs and colorful edible flowers grown on the farmstead.

I grew up in Yorkshire, England. When I was ten, we moved to a small hamlet with only ten houses. Although we weren't farmers, we were surrounded by farms. All my jobs were outside, pretty much. I liked being outside working with my body. I left school at sixteen, which is the age you can leave school in England. I didn't choose to go into further education.

I got a position on an around-the-world expedition called Operation Rally. I met people from all over the world, and I never really went back to England. I ended up buying a ticket to America to work at a YMCA camp. They placed me on Orcas Island in Washington State. I arrived in New York, then I came over to Seattle, got on this little minibus, and I never went home.

I ended up getting a job on a square-rigged tall ship. It was a replica of Sir Francis Drake's ship, The Golden Hind. I fit right in, because I had the accent and I was English, and plus I was invincible, you know. I was nineteen or something. So I worked on that for a year, and I met friends. We ended up jumping ship and traveling around, and I ended up coming down the coast, over the Golden Gate Bridge, down to Santa Cruz. But on the way, I saw the lighthouse youth hostel [Pigeon Point Hostel], and we pulled in there for the night, and I met a man called Three Finger Bill, who was playing his accordion, and followed the music. He brought me to town. I had soup at Duarte's [Tavern] and that's where I ended up meeting my future husband. I was having soup at the end of the bar, and he walked by me, and I said to Three Finger Bill, "*Who* is that?" [laughs] and he said, "That's Tim Duarte," and I'm like, "Okay, great." He was married at the time, but a couple of years later, he wasn't, so that worked out for us. We've been married for eighteen years now.

The landscape around Pescadero is very English. It's very rural. It's got a village life. People know each other. There's a community spirit of volunteerism, connection to the earth. There's a reason people choose to live in a place like this, and I connect with that reason. We all get along very well. It's like an extended family, and it's a feeling of a certain commitment to a lifestyle, which is nice to live around and be around. The hills and the greenery, especially after the rainy season. And the space. It's got lots of good fresh air.

This nine-acre farm was originally built in 1910 by a couple of brothers from Portugal, the Gularte brothers. Frank Gularte was the brother that lived in our house. They operated it as a cow dairy farm for about fifty, forty years. There was evidence of cheese-making and butter-making, with the equipment that we found as we were cleaning out the places, but they mainly delivered fresh milk from the cows. There were a lot of dairies in Pescadero alone. It was a very big industry back then. They would distribute it locally, and then into Half Moon Bay, Santa Cruz. And then a co-op opened in San Mateo, and all the dairies would pool all their milk and ship it over there.

And then, [California's dairy industry] ended up being in the Central Valley. A lot of small family farms started to close in the late forties, the early fifties. And that's what happened to this farm. Frank didn't have any children, so there was nobody to carry it on. I can't remember when the last dairy closed on the coast, but it's been at least forty years. We're the only dairy of any kind in the entire county.

So seventeen years ago, I got the first six goats, but I didn't know anything. I didn't know what I was getting into. It wasn't a business plan. It wasn't something I was on track for. It's been difficult, because I've had to learn everything by myself. I've had no help. There haven't been any resources at my fingertips. But then on the other hand, it's really good, because I've learnt everything very deeply. I've learned how to be creative. I've learned how to do things very efficiently. I've had to think outside of the box on many occasions.

We're a farmstead dairy, which means that we only make cheese from the milk we've produced ourselves from our own animals, so we're not having milk trucks coming in and dumping off their milk. We have to look after the animals from birth to death—from the food, to trimming their hooves, to taking their horns off, to vaccinations, to fencing, water—it goes on and on. That's one huge part of the business, which is very expensive, just by the fact that we're geographically far away from feed lots, you know, grain makers.

Having the cheese-making, having the value-added product from that milk has been integral to any kind of success that we have ever had. So we had to put an incredible amount of money into the farm to renovate it so we could house a dairy, house a cheese-making room. The cheese was incredibly successful, and as the farm became alive again from being a dead, horrible, lifeless place—people were driving by seeing this thing going on, and then there was attraction to it, and it is now such a huge part of our viability, the fact that people come by and buy cheese from us.

When I came to Pescadero, I ended up working at Jacobs Farm, which is an organic herb farm about two miles up the road. I worked there for seven or eight years before starting the farm here. I was there when that business was quite small. I was very fortunate to be involved in a growing business. I got to see all the pitfalls, all the successes, all the hard times, all the good times, getting bank loans, whatever. I didn't even know I was learning all of that stuff when I was in it. We worked really hard getting that business going. And [then] Larry and Sandy Jacobs just kind of left me, went off to Mexico, and were setting up this other aspect of their business, and so I again was forced to learn it. It was like me or nobody, at the time. It was an incredible gift to have that, when I look back.

I was able to learn Spanish, because I had to speak to people over the phone, which [was] really hard. So I got this ability to at least communicate with people. That was an incredible asset that I came away with. And then the business grew, grew, grew, and then more people came to work there, and then I ended up getting married to Tim, and then we ended up having a baby, and I still worked there part time.

But during that time, I sold dried tomatoes to a woman, Nancy Gaffney, who lived in Davenport, who had maybe twenty or thirty goats. She was making cheese. She came to pick up her order at our house one day. I remember sitting on the front steps, and she said, "What a great place. Why don't you buy some goats from me and you can provide me with the winter milk?" Well, that's a good idea. [laughs] I knew nothing. I hadn't had animals growing up. We had like a goldfish, you know, but nothing, because my dad was allergic to fur. But it was a good idea. I used to look out the window, and it was just such a waste of space. There were beans and artichokes being grown in the field, which was fine, but all the corrals. A couple of friends had horses there for a while, but it was annoying that there was just nothing going on.

So I got these six goats, and I think a week, maybe two weeks later, Tim and I went to Guatemala for nine weeks to language school. Larry and Sandy sent me there to learn Spanish so that it would be easier for me at the farm. When I came back, there were only four goats. A couple had died. Somebody was house-sitting for us, and obviously something happened. So from these four goats, I learned how to dehorn them; I learned how to trim their hooves. Nancy, the woman I bought them from, taught me absolutely everything.

Then we had to go choose a male to breed with them. So these four goats had babies. All of a sudden, there's eight goats, ten goats. Those were then bred, and then there was all of a sudden fifteen. So then I start milking them. All the while, she's down in Davenport with her twenty, thirty goats making cheese and selling them into Santa Cruz stores and at the farmers' markets.

Once I started milking the goats, I had to actually store it. So then one of the original outbuildings, which we assume was some kind of cheese-making room or something, because it had a concrete floor, we renovated it. And a woman who worked at Duarte's loaned me three thousand dollars. I put a certified little room. The state inspector came, and he said I needed to do washable walls, and a sink and a fridge and things [like that]. That's how I set up my original dairy, which was in basically a twenty-by-ten room. I had a cold fridge, and I would milk the goats by hand, cart the buckets over, wash all the buckets up and store it. I had a friend who worked at UC Santa Cruz, and he'd put it in the back of his car, and he'd drop it off at [Nancy Gaffney's] house on his way to work.

So originally it was, like, five gallons. Then it was ten gallons. Then it was twenty gallons. Then as my herd grew, I had to grow with it. Then the roofs started getting replaced, and then I had to plant pasture. Then we needed a perimeter fence. I basically reacted all of the time to the growth of the herd. I still didn't know what I was doing. I didn't have a plan, really. It was just, "Oh, I've got to keep going with this."

I got up to thirty-two goats that I was personally milking by hand twice a day. I would milk them, and I'd wash everything up, and it would be time to milk them again. I got a part-time person to come and help, and I got a little portable milking machine, and I was able to milk two at a time. But at this point, I'm producing quite a lot of milk, and it's becoming overwhelming for Nancy, and we start talking about maybe her not doing it any more and moving the entire operation up here.

At this point I've had a child. (Ben's now fifteen and a half.) So I was at that cusp of, "I'm going to either do it or not do it." And Ben was a baby, and I wanted to stay at home, so it was a way that I could still work and stay at home with him. Little did I know it would be way more work than actually going to work, but it wasn't the point at the time. That was the decision I made.

I was able to get a loan because I was a woman in agriculture. I got a low-interest, $50,000 loan from California Coastal Rural Development Corporation in Salinas, and I started developing: putting the silos in for the grain, major fencing work, beginning to think about doing the lower half of the hay barn into the dairy, thinking about a truck, a tractor, an employee. We've moved all [Nancy's] goats up here by this time, because we have the pasture. Her children are getting older. They're thinking about college. She's getting tired. It's too overwhelming, and do I want to buy that portion out?

So basically I was providing all the milk, and she was making the cheese, and then we decided upon a contract where I would purchase her cheese business from her. She'd worked hard. She'd come up with some amazing cheeses. She had an incredible standard, and I was really fortunate that I was in the right place at the right time with the right person. Our relationship was seamless, and the timing was perfect, and it was just this natural progression that was perfect for her, because she wanted to get out, and it was perfect for me, because I had the energy to get in. So she stayed with me for a year and came up and worked two or three days a week teaching me the cheese, by which time, I had got two employees.

[On the farm] there's a hum and a rhythm and a flow on a daily basis, on a weekly and monthly and an annual basis. And it can change any moment; there's always the variable with farming. There's this hum. It starts at half past five in the morning when Roberto and Luis's truck pulls in. The doors slam, and in they go. The lights go on, then the hum of the milking machine goes on. Then you hear the clippety-clops of the goat hooves going into the holding pen, and then there's the milking for an hour and a half. The recycling system goes on, so you hear the fans. And then the dawn breaks. It's light outside. Chickens are clucking away.

One office person arrives. Then all the cheese makers show up—there's three of them—with their little happy chatter, cackling away. And then it's the pressure washer cleaning up the holding pen. And then it's the goats going out into the pasture about nine o'clock, and they're all going out and running around, because they're so happy to be out in the grass.

Then the rest of the people show up for work, and then there's cheese- making. The pasteurizer goes on. The pasteurizer gets cooled. The people show up to get cheese in the shop. The phone's ringing all day. Then there's this quiet time for maybe an hour and a half. And then the night milkers show up. And then it's like the same process into the evening until maybe eight o'clock. And then it's quiet until the next day.

So that goes on *every* single day. And you don't even realize you're in a routine, because you're so in it. You're just a cog in that wheel. That goes from the middle of February to the middle of November, and then all of a sudden, it just stops. The goats are pregnant. We decide to dry them up, and then they have a good two months off. So the clinkety-clankety and the clip-clops—all of those noises just go away. Then something else happens. It's just really busy in the cheese making, because it's Christmas and all the orders are in there, and then there's all these smells of herbs, and flowers are everywhere, and the UPS truck comes in picking up tons of boxes. (We freeze the curd in spring so then we can defrost it at the busy time, which is great, because they don't have four hours of making cheese every day. They can just donate it to packing the cheese to sell.) We're putting Christmas trees up. There's just this other vibe of activity around that time.

But then there's this calm in January when it's kind of rainy, and all the goats are inside and they're heavily pregnant. There's this serenity around the place. Everybody is rested. And then the first baby's born, and it's so exciting. Then the next baby's born, and it's like, "Oh, here, we go." Then it's like blup, blup, blup, and all of a sudden, there're two hundred babies. All of a sudden, one day, the milk machine goes on, and it all starts again. It happens really overnight like that. It's the beginning of our season, and it's exciting.

Then the first three weeks are really tough. The babies are born. They're four days old. They're a week old. The mothers are being milked, but we have to milk them by hand for four days. Then the milking machine goes on. Then the pasteurizer goes on, and of course, a pump breaks, or whatever happens. Then it doesn't stop for another ten months. Then the babies grow up. We keep babies. We sell babies. We retire old goats.

If you don't want to live like this, it would be work. But to me, it's not work, ever. Sometimes when I'm running the business part, the financial part can be very tough. But the actual life is priceless.

Everybody that works here is local to Pescadero. That's been a very resolved part of how I wanted to run the business, was that we needed to sustain not only ourselves, but the community. Four of the people that work here actually live on the farm as well. We have housing for them, which I think is a really important part. When you work so hard, you have to go home to a nice environment so you can rest well, live well, and be able to come back and perform well. So we've invested a lot in that.

Right now, we don't have health insurance. We've had one person that has needed help with her arm, and she's been going through a difficult time. So we've paid for all of her chiropractic, massage, things like that for the last seven months. We look after people in that way. One person had to buy a new truck, so we helped him with a few of the payments. We do it on an individual basis.

I'm lucky enough to go on my husband's health insurance. And that's another thing. He has a very good job. We're very fortunate that he has a regular income. Without that, it would have been very difficult to have got to this point in the farming life.

I don't need to certify ourselves. I know personally how we farm. I know we don't use growth hormones. The cost involved with organic feed is very high. As we saw last year, the cost of feed doubled. We weren't even expecting that. It doubled for the same amount of feed. That is crippling to a small business. But what we've chosen to do instead is people are allowed to come on our farm. They're allowed to come and hear from the horse's mouth how we do it, and they can make their own choice. They can see how sustainable we are. They can see how we recycle rainwater. They can see how we use our own compost that we make from goat manure. For our situation, [certification is] not necessarily going to bring in more income, or make us better than what we are at this point anyway. I'm not closed to it, but at this point, it's not a big issue for me.

We're in control of how we sell the cheese, and how fresh the cheese is, and who we're selling it to. If you come to me today and you want a piece of cheese, I know it was just made just yesterday, and it was milked two days before that, and I know that they've been on grass, and I know that they've had the grain, and I know how fresh the milk is and it hasn't been sitting around for two days, and I know that the chives and the parsley were grown in my garden, and I know that that was done organically.

We are selling 80 percent of our product through the shop and online sales now. It was very, very different even three years ago. We had distributors in Boston, for the East Coast; Chicago, for the Midwest; L.A., San Francisco. We were selling thirty tons of cheese through distributors and 20 percent to our local retail shops, deliveries. I had delivery trucks going out three days a week. I'd diluted myself to the point of losing the business.

One day I woke up and said, "I don't need to do this any more." What I was doing was I was reacting to the phone calls, "I want cheese; I want cheese." "Oh, we've got to make more cheese. We've got to keep more goats. We've got to milk more goats. We've got to get more people to milk the goats." We had 350 goats at one point, and it was awful. It was beyond capacity for the people that worked here, the machinery. The goats were compromised. There was no space for them, and it was just over. One day, I woke up: "It's over." I sold 150 of my goats to the rent-a-goat program.

It was the best decision I'd ever made, because at that moment, I realized I was in control, that my decision controlled how I wanted to run my business. I didn't know that I had that control before then. But I know [now] that I'm not going to compromise the health of the animals by stuffing more animals in there. I know I'm not going to compromise the team of seven people that we have by diluting them with more. I know that I'm not going to overrun the building, because it's not bigger than what it is. This is what works for us. I have to really look at my life with my husband and my son and see the impact that that has on my family, and none of those are compromised any more, ever.

So we decided to come out of all of the stores, because we could see the rise in the [on-farm] shop. We could see the rise of online sales. We knew that we would run out of product by Christmas. We just knew it. Plus, I had less goats, because we'd cleaned out the herd a little bit. So I was making a little bit less. It was like we had to do it or we were going to be completely screwed. [laughs] We basically got rid of 80 percent of our income in one day.

And it was very scary, because my payroll was the same. I wasn't going to not have people work. My bills are the same. My mortgage is the same. I still have to feed all the animals. I just have to say that I have a very amazing bank that I work with. I've had a ten-year relationship with them. I have a daily relationship with them, and if we didn't have them, I wouldn't be in business. I consider them a partner. They're our local bank, First National Bank here in Pescadero. I know everybody from the top to the bottom, and it's an absolute partnership.

But even [with a] good relationship with a bank, it was still very difficult. People had to go on reduced wages. Fortunately, I own the house where the farm people live so I just reduced their rent to the bare minimum to pay for the mortgage on it. Two of the people I kept at the same wage, and they just took on a bit more responsibility. It just unbelievably worked out. And then Christmas came, and we had product in the freezer, and we were able to sell it all. We made profit on it, so we were able to pay for at least a couple of months that were really tough.

And then going into January [2009], this whole economic thing happened. The world changed, and we benefited from it. From January 1 to now, we have done better than we've ever done in our entire existence, because people were, "Oh, let's go to the farm, because we want to feel good." So now everybody's back to work. Everybody came back to work. Everybody's working harder than they've ever done before. [laughs] We're actually hiring people.

It's a thrill, and you realize out of what you consider bad things—it's not even bad things; tough times—you do prevail, but you have got to be able to get up in the morning. It's not for the faint of heart, and it's not about getting rich. You have to feel the richness in your day-to-day life.

The other thing that we're huge into is the ecotourism. There was no name associated with it, but when we first had our first baby goats seventeen years ago, the local schools came up. The little kindergartners would totter up, and they'd come and play with the baby goats. So we've been doing that for seventeen years. Now it's turned into a major part of our viability.

As the farm became more of a desirable place, we painted the barn. It was so idyllic and beautiful. People would drive by, and then they'd pull in, and then they'd knock on my door. I'd be having my dinner. They'd be coming in the back door. They'd be coming in the front door. I'm like, "What? Who are you?"

"Oh, I just want to see the goats. I want to buy some cheeses." Oh, my God. So we opened a little shop so people could buy cheese. The original birthing stalls in the lower half of the hay barn, we had it as a tool shed for about ten years, and that's what we decided to use as the shop. And it just started with a five-by-five [foot] thing, and we put some cheese in there and a cooler. Then it expanded from there, to the point where it's almost beyond capacity on the weekends. It's heaving with people. It's so gorgeous. It's the original redwood floors, and all the original redwood beams are there. They're all undulating where the cows used to champ around and probably nosh on the beams and things.

So we did a lot of school groups for many years, and then we started adding adult tours, where people paid to come on [the farm]. Now we are completely booked months in advance. We take twenty adults. We do them eleven o'clock and one o'clock on Saturday and the same on Sunday. We've now added a Friday. People pay, and they get two hours of our time looking around the farm. They go from the very beginning to the very end, and then they taste the cheese.

We have five tour guides now. They're all local people who live in Pescadero. One woman is the schoolteacher. Another woman works with special-needs students. We picked the people, asked them to come and work at the farm. They don't need to be theatrical and they don't need to be a showman. They need to just impart their story. "Why do you live here?" is a big question. "What's the school like?" It's a real connection with somebody, and why they want to work here and do this. And they love it, because it's completely different to their regular job throughout the week. They're paid very well, because that's our marketing. We pay them to market our farm. We don't advertise, never have done, and we have done it from the inside out.

The school groups, they book a year in advance to come now. It's like every third grade goes to Harley Farms in April, or whatever it is. So we're so ahead of that, that actually the tours don't interrupt the flow of the farm.

We've had dead babies out there. It's an opportunity to say, "Yes, it's a dead baby, and this is probably what happened." I had this terrible birth in the middle of a tour. I could not get this baby out. It was dead. It was huge. The mother was bleating and having a hard time, and these little girls, they must have been four years old, they're on the tour, and they couldn't stop watching. And the parents were having kind of a bit of a hard time, "I don't want my kid to see this thing."

Anyway, these two little girls came over. I'm having a little break, wondering what I'm going to do. I've got this goat half hanging out, and this one little girl said, "Excuse me," and she said, "My friend wants to ask you a question." [laughs] I said, "Oh, yes." She said, "Is that baby dead?" and I said, "Yes, the baby's dead, and it's really important that I get the baby out so the mother will be okay. That's all I'm concerned about now, because she needs to be okay."

At which point, the mother and father came up behind, and they were really afraid, but we had this really nice conversation on a very basic level with these two little girls about death, and the fact that what was important, right then and there, was this mother. And they went, "Oh, thank you," and off they tottered. They just needed to know the truth. And I said, "You're going to go in [to the loafing barn] right now, and you're going to see two hundred babies, and you're going to have so much fun." "Okay, bye." And then the parents went, "Oh, thank you. Thank you."

We have the opportunity to show something that's real and not be afraid of it. Things like this happen all the time, and it's pointless sheltering people, because then you're sheltering them from life. I would say everybody on this farm is capable of having that conversation with anybody. "How are things going on the farm?" "They're not good." I mean, what's wrong with that? [laughs] It's okay. We're not so idyllic that things aren't not good sometimes.

I keep bringing "real" up, because I think it's important that people are not idealistic about it. You have to be realistic. Goats are going to die. Goats are going to get sick. You're going to have to make some tough choices. You're going to have to put your finances on the line. You're going to have to fire people if they're poisonous to a team. And it's not, again, for the faint of heart.

I think that every business in our village contributes to our success, because we're all talking about each other, so we're all helping each other funnel people to our stores. We have a really good working relationship with all of the business owners here. They're talking about how proud they are of this farm, and how they've seen it grow from nothing. So they're all marketing in a way, subtly, and they're using our product to take to places, because it's this sense of local pride. They're really the backbone of our success. I mean, we're in a beautiful place, too. That helps as well. But they're on our side, so it's important to reciprocate.

Our barn is used constantly by local organizations for meetings. We do an annual cheese-and-wine event up there. We donate pounds and pounds of cheese to whatever anybody wants. We're very involved with the school. I sit on two boards in the community. I have done that for many years.

And then, we're an avenue for work for the high school students. We have six. Two are going to graduate this year. They've worked for three and a half years for us. It's their first job. We're not easy with them. We don't let them get off with stuff. When they leave here, they know how to work, and they know how to take responsibility. We treat them like adults. They're a big part of the fun of doing the shop and the farm. I love having those guys around.

I do the local Pescadero Arts and Fun Festival. I'm the emcee for that. I'm kind of an advocate for the town. We get a lot of press. We're very fortunate. I always mention the town. The town is paramount. It just is.

Cynthia Sandberg

Photo: Tana Butler

Cynthia Sandberg is proprietor of Love Apple Farm, an establishment unique among Central Coast small farms in its combination of biodynamic techniques, an exclusive supply relationship with a single high-end restaurant, a focus on heirloom tomatoes, a rich public offering of on-farm classes, and a successful Internet-based marketing strategy. As the kitchen garden for upscale Manresa in nearby Los Gatos, Love Apple enjoys a symbiotic business relationship with the two-Michelin-star restaurant and its executive chef-proprietor, David Kinch, who often visits the farm. While Sandberg grows a wide variety of produce for Manresa and for sale in her seasonal on-site farm cart, she specializes in heirloom tomatoes. She also teaches popular classes on a wide variety of topics including cheese-making, home canning, nontoxic gopher control, and beekeeping. From her humble beginnings in Ben Lomond, Sandberg moved her farm onto a twenty-two-acre site in Santa Cruz where she is expanding her curriculum to include more gardening, preserving, and cooking classes.

I was raised on the east side of San Jose, California, right next to a pear orchard. They eventually tore the orchard down to make way for a school and all the urban sprawl that was happening in Silicon Valley at that time, taking out all those great fruit orchards and putting in housing and schools, but I remember playing in the pear orchard and in walnut orchards nearby.

I eventually decided that I'd become an attorney. I went through law school, and I was practicing law for a while, bought one of my first houses with my husband at the time, in Capitola, California.

I started to garden there and wondered the first year why some of the plants that I had put into the ground didn't survive the winter—because we have mild winters, particularly near the coast. And I thought, well, I'm putting all this time and effort into my garden. I'm going to go learn about this a little bit more. So I started taking horticulture classes at Cabrillo College. I found out that there was this thing called an annual. And that's why some of my plants didn't make it through the winter. I had no idea, starting out as a new gardener, that there was even a difference that basic.

Cabrillo College changed everything. It's a wonderful horticulture program up there. Richard Merrill was quite a force in the horticulture field in Santa Cruz County.[1] He was a long-time professor and probably the driving force behind the whole program. I went through the program, and right at the end of it my ex-husband and I bought this farm. It wasn't a farm at the time; it was just a two-acre piece of land with an old farmhouse on it, up in the Santa Cruz Mountains. It was around that same time that I quit my job to focus on raising our small son.

And since I can't just sit around and not do anything other than mothering, though I love being a mom, I started ripping out lawns and putting in vegetable gardens, and every subsequent year, more of those bits were devoted to tomatoes. That was what really became my passion. All different shapes and colors and tastes and sizes and intricacies go into heirloom tomatoes. Every color of the rainbow, you can get in an heirloom tomato. So I wanted to grow this one and grow that one, and oh, I have to grow this one too, until every nook and cranny of the property were covered in tomatoes every year. Then I'd take a break in the wintertime and start looking at seed catalogs and getting familiar with the new varieties and new techniques to master tomato growing.

The way I started my little cottage business ten years ago, I wanted to grow—I remember this very distinctly—ten varieties of tomatoes. I laugh at that now, because I'll grow 125 this year, but at that point I thought, wow, ten! I bought ten seed packets, and I had really good luck germinating them, and all of a sudden I was faced with 300 healthy, fabulous tomato plants. What to do with them? Well, I live on kind of a busy road. I put a sign at the end of my driveway,

---

[1] See the excerpt of the oral history with Richard Merrill in this book.

and I said, "Tomato Starts, a Buck," and people came and loved it. I had this idea: Aha! I could pay for my seeds every year, and my little tomato hobby, if I sell the extra tomato plants in the spring. And I can make a little bit of extra money selling all this extra tomato fruit that I'd get in the late summer and fall. I had a fruit stand loaded up with tomatoes, right in the driveway. And people started to come.

It was on the honor system back in those days. I was too busy working in the garden to interact with all the people that wanted me to weigh produce or take change. I didn't have to make a living at it, because my husband still went off to work and made good money and supported my habit. I still tried to make some money so that I wouldn't feel so bad about expending this huge sum on soil and trays and pots and seeds and garden beds and soil amendments and everything that I was doing, trying to grow as good a tomato as I could get.

Eventually I became totally tomatoes, and since I was growing a hundred varieties here on the farm and offering them for sale as starts in the spring and on the tomato stand later, I started getting noticed by various media: What is this freaky tomato lady doing and why is she doing it? First off, newspapers would come and do a little article, maybe in the garden section. And after that, some magazines. I got into *Sunset* magazine. In California, the pinnacle of every gardener's dream is to be in *Sunset* magazine, so that was good fun. And along about that same time, I was featured in a couple of different television shows.

That was about, I guess, the pinnacle of the tomato craziness, about four years ago. After that, my ex-husband kind of gave me an ultimatum. He says, "You're working yourself to death. I kind of want my Martha Stewart wife back that I used to have," and I said, "Well, I really don't want to stop Love Apple Farm. I created it out of my own sweat, blood, and tears, and I've got thousands of customers that come onto the property and tour around." But it was getting to be too much for him, the lack of privacy. I could see where he was coming from, and I empathized with that, but I said, "You know what? If that's what you want and this is what I want, then I agree with you. We'll part company." And it was a mutual thing.

In the divorce settlement, I got the farm, but I also got the big fat mortgage on the farm. I had to make a decision: okay, now I have to really support myself. It was no longer a hobby. I have to make the farm pay off. So I started giving gardening classes. I had so many people asking me, "How do you grow these tomatoes? How do you do this? How do you do that?" Pretty soon I'd attract a little crowd of people that were coming to the tomato fruit stand or to get some seedlings. I thought, Well, I'm going to see if I can actually charge for a gardening class.

Now, four years later, that has become a huge part of my business. I started giving tomato gardening classes, and tomato seed-sowing classes, and summer vegetable gardening, and winter vegetable gardening. Now we're up to jam-making classes and beekeeping and compost and ver-miculture. It's been a nice way to supplement the farm income.

Another way was my partnership with Manresa restaurant in Los Gatos. The chef, David Kinch, had heard about me. David Kinch loves to source local produce, meats and fowl, fish. Somebody kept telling him about this crazy tomato lady, and he asked me if I would start supplying him with tomatoes. I'd had that request before, from other restaurants, but I always turned them down because it was all I could do, really, to produce the tomato the way I like to grow them, which is very labor intensive. It was all I could do to put them on the tomato stand and manage the farm by myself, so I usually turned down those requests from chefs because I can't be putting the tomatoes in my truck and schlepping them around to one county, let alone another county over, which is where Manresa is.

But I said yes to Chef Kinch because his restaurant had recently been named one of the top fifty restaurants in the world. I thought, wow, that's pretty posh! I'm going to go check that out. So I had dinner at the restaurant. It was a life-changing experience. It was the most wonderful thing to eat there, and I agreed to supply him with tomatoes for a season. And the season ended in late October, like it does here in Central California, and that was that. I went back to scraping by and trying to do some other things. He didn't get all my tomatoes anyway that year. I still had my tomato stand.

But the following spring, he called me, and he said, "I'm looking for my own farm." What he didn't see at the other end of the line was me raising my hand, like, "Me, me, me, me, me! I want to be that. I want to be that farm." Because I needed to have another revenue source, and I also knew what the chef's cuisine was like, and I told him right away—I said, "It can't be by the pound, because you give such care with your food and you have such small little bits and pieces of all these different things on your plate. I know you want borage flowers, for example. I'm not going to charge you by the pound for borage flowers that weigh one-eighth of an ounce." So it would have to be something like a community supported agriculture chef, with one restaurant. I joked that it's an RSA [restaurant supported agriculture] instead of a CSA.

So we struck a deal for a monthly amount, and we started out very small, just to see how it would go, the relationship. Now, three years later, almost all the things we do here on the farm go towards the restaurant. All of our planting beds are devoted toward creating something for Manresa, not just tomatoes. I got back full swing into every kind of fruit and veg you could imagine. I can't grow everything up here in these cold mountains, but I can grow a lot of stuff that you can't grow down in Santa Cruz proper, near the coast. I can grow stone fruits here, where they can't. They need x number of chill hours every year. What's nice is in every garden, no matter where you're located, there's a niche for you.

Growing for the restaurant is a constant challenge because I have to produce something for him fifty-two weeks of the year, and it has to be two-Michelin-star level, and we're shooting for three Michelin stars. He's renowned for his vegetables, and I have to uphold that reputation. So it's always a challenge trying to make sure that I give him enough different things, and new things that he can't get anywhere else, because it's got to be a mutually beneficial relationship. He's got to see some value in having his own farm. It gives me challenges that I wouldn't find in another type of a farm that has to be more consumer-driven.

I'm always trying to shake it up: new varieties, new things that he's never seen before. Sometimes it's not successful, and other times it's fabulously successful. Whenever I get something new that he really likes, he gives me this statement, "This is the single best thing we grow here." Then three months later it'll be something new, and he'll go, "This is the single best thing we grow here." [laughs] And it's just become kind of a running joke for me, to see if I can get him to say that yet again about something different.

It's a very collaborative relationship. He does tons of research himself. He goes to a lot of seminars. He goes to restaurants all over the world. That's part of his job, to find out what's new and exciting and improve on it, make something new that nobody else has made before. He'll collect seeds for me in other countries and make sure that I get my hands on them. I do a lot of research at night, whether it's on pest or disease control or new varieties. If I see something online that looks intriguing, I'll acquire the seed and I'll put a little test patch in the garden. Sometimes I won't even tell the chef that I've got something new going until it's ready. I'll say, "Taste this. See what you think." Then it's, "This is the single best thing we're growing."

He's very involved. He loves to walk through the garden. He's always tasting and smelling and touching and thinking. I can just see the wheels turning in his head about recipes that he might make out of this or that. He'll even take things from our pine and fir and juniper trees to make gelées and infusions. He'll go foraging in our tiny little stand of redwoods. Sometimes he'll pull out a weed or a mushroom that he knows is edible, and that night one of our weeds is on a plate at Manresa restaurant, being served to the *Michelin Guide* director.

There's an annual tomato dinner, where everything on the menu has a tomato in it, even the dessert. He gives me a table, usually for six, and I bring my really good volunteers and interns. I run on a system here of volunteers, interns, and apprentices. People love to come in and learn, and I'm happy to put them to work. I reward them by taking them to the tomato dinner or taking them at other times to Manresa. And it's a reward for me, too, because it really regenerates my batteries and makes me want to do even more stuff for the restaurant.

Our deal is to have this be a productive farm year-round. They're open year-round, and they need produce year-round. The chef is very seasonal in what he likes to put on plates and his diners are very savvy. They don't want to see a tomato in January or April. So you better be having the tomato in the right time of year. You better be having the broccoli in the right time of year. We can only do so much as far as the seasons, but we still have maybe a hundred different varieties of things growing at the height of winter, and at the height of summer, maybe 200, 250 different things growing for the restaurant.

What is on the menu is driven by the garden, but I can't produce everything for him. I can't produce citrus or avocado. My climate doesn't allow me to do that, and being such a haute cuisine restaurant, he's got to have citrus in there, and he's got to have avocados. So we try our best. I think when we first started it was 30 percent, and then it got up to 50 percent. I think now he's running at about 80 percent. And he tries really hard to ensure that everything that we produce, he uses.

When we first started out over three years ago, it was a new business model for farmers. We got a lot of press because of that interesting relationship. I know Blue Hill at Stone Barns, above Manhattan in New York, they have their own farm. But it's kind of a different relationship because that farm is supported by grant monies and—I could be wrong, but I heard Rockefeller money. And when you go there, it is a wonderful place, but it's not a farm that has to make a living doing what it's doing. *I* have to make a living doing what I'm doing, so there's always a mind towards: Is this paying off? Am I saving money? Am I saving energy? Am I able to buy that bag of feed for the chickens? Can I produce enough compost and worm castings here so that I don't have to go buy my fertility?

Even before I had met the chef, I was at a farm in northern California that grows biodynamically, and I had never heard that word before. This was in October, maybe five years ago. Anybody that grows tomatoes realizes that by October, they've already passed their prime. Ugly-looking tomato plants, usually, in late October. But these tomatoes were not only thriving and producing fruit, they were probably the most beautiful tomato plants I'd ever seen, and I've seen a few tomato plants in my day.

I couldn't ask the farmer that day what was going on, but I ended up seeing him a couple of months later. His name is Jeff Dawson. He's very famous in biodynamic agriculture. To me, he was a rock star. It was like meeting Bruce Springsteen for me, to meet Jeff Dawson, because he's a big tomato guy. I said, "Jeff, how can you get a tomato to look that great in late October?" and he said, "Biodynamics." I said, "What?" He got a look on his face like he'd probably had to explain that to too many people, and he said, "Just Google it." And I go, "Okay." I didn't want to take up too much of his time.

I went back home, and I Googled it, and I realized, Oh, geez, this is kind of strange. This is kind of out there, biodynamics, because you're looking at moon phases, you're looking at cow horns buried in the earth for six months, you're looking at chamomile stuffed into stag bladders and buried in a muck for another six months. If I hadn't have seen those tomatoes growing, I would have completely disregarded it, but the proof was in the health of the plants.

When I was discussing with Chef Kinch the way the farm would shape up for the restaurant, at some point during the conversation he says, "You know, Cynthia, I also want the farm to be biodynamic." I had this expression on my face like, what, what? I didn't even want to mention it because it's too strange. But he says, "No, I see that as a viable option in the future for organic agriculture." It was kismet. We were on the same page at the same time, and we started full-force in on learning about it.

Chef and I decided the farm would be biodynamic from the get-go. One of the tenets of biodynamic agriculture is creating your own fertility and having a closed loop between your inputs and your outputs, so that you don't have very many inputs that are off-site. That necessitates a lot of our own compost and fertility, so we have chickens; we have a pig. We have worm bins. We have compost piles. We get the kitchen scraps from Manresa. Everything goes in the compost or to the chickens. That gets regenerated into the garden. It's a very good, closed-loop system that includes the restaurant. So on that basis alone, I can get behind biodynamic agriculture.

I'm not [certified organic]. I've never sought that out. I'm only one gal. I have a lot of volunteers now, but back in the day, when it could have mattered, I didn't have the time or the money to seek organic certification. I was already working eighteen hours a day trying to make this business go by myself. I couldn't put on top of it all these restrictions and requirements of organic certification.

All my customers come to the farm. You come to the farm to buy your tomatoes. You come to the farm to buy your seedlings. No part of my property, including the interior of my house, is off limits to people, because they have to go into the house to go to the bathroom. So everybody goes everywhere on this property. They see what I do. They see how I do it. They see that there isn't a stack of Roundup® in the corner or a stack of Miracle-Gro® over there. They see the compost piles working. They see the chickens in action. They see the worm bins. They see us working every day. So my customers didn't need that additional validation from a government entity, because they could see it with their own two eyes.

Going to a system of interns and volunteers and apprentices keeps my labor costs down to zero, basically. When I first started the farm, I did everything on the farm, or my husband or my son or my friends helped me. People who were interested in gardening but they didn't have their own patches of land, they'd come over and help me. Even my customers started helping me. They became my volunteers, and that's how we've worked for many years now. I can grab some great farm interns from UC Santa Cruz. They have a fabulous program, and the young people that I get here from that program are very well educated, very passionate about making things grow properly, about pest management, about disease control, all organically done, and they learn a little bit about biodynamic agriculture and get a chance to get their feet wet and their hands dirty. As we speak, there are six people on the farm that are volunteers, turning in cover crops, watering tomatoes in the greenhouse, thinning vegetable beds. I've been really blessed to have that kind of help.

My interns are people that have committed to a certain amount of months and come more often than once a week. They're many times associated with the university or they've made this verbal commitment to me—you know, "I really want to learn how to do this. I don't need housing. I don't need board on the farm, but I really want to learn," and we set up three days a week they'll come over at certain times for six months. Those people, I consider interns.

And then my apprentices—I've just got one apprentice now, and she lives on the farm with me and works and learns. She wants to be a farmer one day, and this has been a fabulous help for me because then I don't have to do all my opening and closing chores like I had to do for years and years here. That was really quite exhausting. I have somebody else that can help me out with that. Because people come, and they go, volunteers and interns, but if you have an apprentice actually living on the farm with you, they can go out at midnight and lock up the chickens that you forgot to lock up earlier. That's very nice.

I found her through advertising at the ATTRA [National Sustainable Agriculture Information Service, formerly Appropriate Technology Transfer for Rural Areas] website. It was a lark one day. I was doing some research on the website, and I noticed they had a page on farm apprentice programs, and I thought, well, I'll put something on there. And since I put that on there, I've had dozens and dozens of inquiries from people wanting to be an apprentice.

I see a marked increase in the amount of people wanting to get into farming and agriculture. I think the fact of my association with the restaurant means that I get more press than a funky, two-acre microfarm would normally get. People want to be part of that world-famous restaurant connection. I'll take it. Anybody who wants to come give me a hand and shovel some compost around, I will take that help. And they get a benefit, too.

I find that the more income streams that I have on the farm and the more media attention I get, the more I need to be on the Internet, ordering seeds, making Excel spreadsheets. I still do all my own accounting. I've got irrigation to install. I haven't taught anybody how to do that yet. I've got class administration to handle, which is quite onerous. I've got to market the classes. I've got to market my tomato seedling sale. It's ironic, because I started the business because I like to get my hands in the soil and I didn't want to be at a desk job anymore, and now I've made the transition of not being able to be outside working as much as I'd like, and I have to sit at the desk again. I'm not sure how I get back outside other than hiring people to do the accounting and do the marketing for me. I'm not quite ready to make that commitment yet, nor is that in the budget. But maybe one of my good interns or volunteers can set me up with that.

When I first started out with my passion for heirloom tomatoes, I was one of the only games in town. You could not buy heirloom tomatoes at Safeway or Costco back in those days. When you went to a farmers' market ten years ago, you maybe got four or five different varieties of heirlooms, maybe a bicolor, maybe a black, like a Cherokee purple, but you didn't have the kind of selection you have now. And that's not because of me. That's because people have reconnected with their love of gardening, their love of heirloom varieties. So I have had nothing to do with it other than realize that, wow, the competition is really increasing there. So I continue to try to adapt and change so that I can still earn a living.

I'll do a lot of research in the wintertime on varieties of tomatoes worldwide that we don't have here yet in California and you can't get at any other nursery. I'm on these tomato boards all over the world, and I converse with other gardeners. If there is a tomato that somebody loves in Iran, I'll acquire that seed somehow. I did one year acquire a fabulous tomato seed from Iran. I got it from a friend of mine who is Iranian. He brought it to me, and he said, "This is the kind of tomato we have in Iran. I don't know the name of it, but here are the seeds. Let's try it." So we brainstormed some names, and I asked him what the word for tomato was in Farsi, and he said some word that was way too long and complicated. So I said, "What's the name for beautiful?" He said, "Ziba." I said, "Well, let's call it Ziba." That was a popular tomato for a few years for me. That's how I adapt, always trying to offer my customers new stuff.

I can be having a really bad week with too many pests, too much heat. It'll be one hundred degrees out there, and we're just working and sweating, and maybe the squash plants haven't produced the way they should, and I'll be thinking about chucking it all in, and I'll have that one customer come in and just go on and on about how great it is and what a change it's made in their life.

I had a gal wander in, and she goes, "I want to thank you. I have lost one hundred pounds." And she showed me an old picture of herself, and she says, "A year ago I came to your tomato stand, and I got your tomatoes, and it inspired me to eat more healthfully, and I have since lost one hundred pounds, and I credit your tomatoes." I thought, This is crazy, people saying that kind of stuff to me, and I'll tell you, that's great for me.

Or the chef at the restaurant will say something about, "We bought some black kale the other day from a farmers' market because we needed it for an event and you didn't have enough for us, and one of the cooks came to me, Cynthia"—this just happened a couple of days ago—"they tasted the dish made with the farmers'-market kale, and they said, 'We can't serve this. It is not up to the standards of the kale that we're getting from Love Apple Farm.'" I don't know why our kale tastes better than their kale. Maybe it was an anomaly. But that kind of comment will bolster me, even on the worst days.

It's very complicated, being a farmer. It's way more complicated and challenging than people think. I've had people come into the garden and make comments like, "Oh, things just grow so easily here. All you got to do is dig that hole and water it, right?" I have to just kind of take a breath, calm myself down, smile, and say something to the effect that it's a little bit more complicated than that. And that's where the gardening classes come in, because a lot of people try and they fail, and they go, Hmm. That's how I started taking classes. I said, this is a little more complicated than digging that hole and Miracle-Gro'ing.

It ain't such a miracle. Where is my miracle? [laughs] I'm in charge of tens of thousands of life forms here, and those life forms depend on me to keep them fed and watered and fertilized and happy and warm at night. Farming has been way harder than being an attorney. A lot more brain power, a lot more effort, a lot more frustration. It is not for the faint of heart or the weak of mind. But I will tell you that the worst day gardening is better than the best day lawyering.

Big ag is never going to be able to come in and do what this farm does for Manresa. Chef likes to take [plants] for every stage of their life cycle, from the tiniest sprout, to the seed. He takes it in every level of its maturity, depending on the plant, and experiments. From cucumber seeds, to lettuce, to newly germinated lavender sprouts, he'll take it all and try to make something edible and wonderful. And as long as you have that niche and you have the ability to do it, then I don't think that a factory farm can come and take that away, because it *is* all about detail.

José J. Montenegro

As a child, José J. Montenegro was troubled by the poverty in rural Mexico. He studied agronomy in 1988, and despite sadness about leaving his homeland, decided to emigrate to the United States. Montenegro became the farm operations director of the organic farming training program at the Rural Development Center (RDC) in Salinas, California. The RDC initiated a 'Farmworker to Farmer' program where agricultural workers received training that allowed for their advancement on the job, in farm management or ownership. This program eventually became the Agriculture & Land-Based Training Association (ALBA). In 2000, Montenegro left the RDC to begin Proyecto de Arraigo, *a program that offers training and resources to farmers in rural Mexico. He recently earned a master's degree in public policy from California State University, Monterey Bay.[1]*

I was born in the state of Durango, Mexico, in the southern part of the state, in a town called Nuevo Ideál. I was just born there, but I actually lived and grew up in a small farming community called Providencia, a minute, adobe town hanging from a magnificent mountain. [My background has] most definitely shaped not only my life and identity, but my professional interests, my perspectives. I was a child of small family farmers in a community where, like many small rural communities in Mexico, we were disengaged from civic life and civic participation. These communities were gradually being displaced.

Paradoxically, I was able to look at that experience once a process I refer to as "forced migration" brought me to the United States, because

---

[1] Montenegro edited this excerpt fairly extensively. The more lengthy additions are included as footnotes. For a verbatim transcription see:
http://library.ucsc.edu/reg-hist/cultiv/montenegro

it made me realize what the barriers were that these communities confronted in achieving sustainability. I was able to see and understand how there were erroneous agrarian reforms and policy-related issues that have influenced my peasant-based community and many other communities in my region in the state of Durango. These communities, for the most part, did not have access to basic resources. I think this impacted the communities' ability to be successful and to thrive and to maintain their way of life.

Over the years, I saw how farmers began to sell their tractors. They began to sell their horses, cows, pigs, and chickens. They also began to sell their lands, and then emigrated and left their communities to sell the only thing they had with them, their labor. Most people did not want to leave their communities of origin but felt that "*la tierra ya no da para*," "the land can no longer support us."

These were isolated rural communities. There wasn't a whole lot of advocacy on behalf of these communities going on, no organizing to represent the rights and the needs and aspirations of these farmers and communities. All of that put together had a major impact. In time, so many of these communities became ghost towns. Young people left, which represented a loss of social capital, of human capital. Unfortunately, young people who leave these communities do not go back. These rural communities were experiencing a new phenomenon, "forced emigration," characterized by dramatically deteriorated living conditions.[2]

I never wanted to leave my community. I told my father especially, and my mother as well, that I didn't want to go to school, that I wanted to stay at the farm because that's what I loved and that's what I was

_____

[2] Montenegro added the following written comment during the editing of this book: Many people from rural Mexico, like me, have emigrated to the United States. From one form of invisibility in Mexico, we entered another world of invisibility in the U.S. As Dr. Joséba Achotegui, a tenured professor and cultural psychiatrist at the University of Barcelona, describes in his paper "Immigrants Living in Extreme Situations: Immigrant Syndrome with Chronic and Multiple Stress (The Ulysses Syndrome)," published in the journal *Norte* of the Spanish Association of Neuropsychiatry (2004), "Immigrants have to behave like heroes in order to survive. Ulysses was a demigod who, despite this, barely survived the terrible adversities and dangers he had to face, but the people who are arriving at our borders today are creatures of flesh and blood who, nevertheless, must suffer episodes that are as dramatic as, or even more so than, those described in the Odyssey." In my community of origin young people lost interest in farming because it isn't viable; this leads to "forced" emigration, a human exodus. So these are part of the trends that impacted my own life.

passionate about and that's what I wanted to do. And my father said, "You know, things are going to become more difficult at the farm level. You should definitely think about other options. If, after you go to school and graduate, you see that there is viability and possibility for you, well, you can come back to the farm."

I said, "Oh, okay. I'll go to school, but I promise you I'm going to come back to the farm." For high school I ended up going to an agricultural-based technical school in the state capital, Durango. From there I went to a local university and earned a degree in agronomy. However, by the time I graduated, in 1988, I looked back at my community and saw that what my father had predicted, his prophecy, had pretty much become a reality—economic conditions for peasant-based communities were so difficult that I saw little hope and little possibility. I thought, while I want to go back to the farm, maybe there is another way for me to get back, for me to contribute to the restoration of a way of life. Maybe there is another way to fight back.

The government system at that time was so corrupted and so disengaged from people's lives and the reality going on in these communities. I wasn't going to be a part of it. While I had opportunities to work for the government, I felt that I was going to be betraying my own way of life and my father. [I saw] the profound disengagement between government institutions and rural communities. There was a major misalignment between the programs and services offered by government institutions and the needs of rural communities. They did not match. That misalignment has led to major gaps and problems. I said, I don't want to be part of it. I wanted to "escape" and just go away. [I] ended up coming to the United States. My choice was to emigrate. While emigration represented a dramatic episode in my life, and involved loneliness and forced separation from my loved ones, immigration into the U.S. also offered me new opportunities I never thought were possible, including living in highly diverse, multicultural communities.

I was hired as the farm manager-educator [at the Rural Development Center in Salinas, California]. There were two roles or responsibilities. One was to manage the farm—things like taking care of the irrigation system and the land, planting cover crops, maintaining the equipment, etcetera. But also I was responsible for providing resources and technical assistance to the program participants, including teaching some of the courses.

When I got there, the education curriculum that was in place was rather informal, and so Ann Baier, who at that time was the director, and [I] had an opportunity to look at the informal program that was in place back then and propose a more structured course, which later became known as the Small Farm Education Program course, or PEPA [*Programa Educativo para Pequeños Agricultores*], a six-month, intensive course that is required for the program participants in order for them to have access to the land at the farm, and machinery, and irrigation equipment. Ann, I, and others, including the program participants, worked in formalizing this six-month training curriculum. We included in this program curriculum topics around sustainability and organic agriculture.

The Rural Development Center was sort of a wonderful model and experiment, if you will, in that, initially, back in 1993, most of the land at the Rural Development Center was farmed conventionally. And in the course of five to seven years, close to 90 percent of the land was transitioned into organic certified land. A major, major transition was underway [on this] hundred acres of farmable land. But that was the result of an evolving process. It was an organic process that took place, and it was a process of ongoing dialogue with the program participants, mainly immigrants from Mexico. There were many challenges along the way. There was some resistance to going to organic agriculture, from the program participants, primarily. The resistance [was] coming out of not having experienced and seen the benefits from organic agriculture before. Sometimes you don't know what you don't know. Many of the program participants came from farming communities in Mexico that were largely dependent on pesticides and synthetic fertilizers inherited from the Green Revolution in Mexico and many parts of the world.[3] They viewed the use of chemicals and pesticides as the most viable solution, because that's what the institutional system preached and promoted among farmers in our regions.

And so, just to give you one example, let's look at the farm as an ecosystem. That includes planting windbreak trees, diversified trees to protect your soil from erosion, environmentally sound soil management practices, and so on. A lot of the program participants expressed

---

[3] The Green Revolution began post-World War II, when plant breeder and soil scientist Norman Borlaug spearheaded a Rockefeller Foundation-funded program to increase yields of wheat in Mexico. The term itself refers to the attempt to increase crop yields through irrigation, "genetically improved" hybrid seeds, petrochemical fertilizers and pesticides, and mechanization.

concern. They said, "Well, you know, the trees are going to provide a habitat for birds, and the birds are going to eat my seeds." Some of the concerns may be legitimate, but we had to sit down and look at the cost/benefits of things. And rather than planting one thousand trees at once, we had to plant a few at once and allow the program participants themselves to see the benefits. The use of cover crops—that was another example, just allocating a small section of their plots to try out different cover crops and then seeing the benefits the following year as their crop yields improve and so the quality of their products.

So it was a gradual transformation. The challenge was more in the area of attitudes and behaviors. How do we change attitudes and behaviors? A lot of dialogue among farmers and information and education had to take place. It wouldn't work had our approach been paternalistic, top-down. It was a difficult process of ongoing dialogue, and it was a learning process for all of us. It started as a result of the farmers themselves expressing interest in turning the Rural Development Center into an environment that was safe for their own families. No hazardous pesticides. That's how it started. That was the spark. "I'd like to see my child come to my farm and help out but also see and learn. I'd like to see my wife being a part of the operation. And we'd like to see this as a community space." So that was a major turn along the way. As a result of that, we turned to organic agriculture. We started with four organic acres, experimented with them, and then after a couple of years, our student-farmers got into it.

They also obviously saw the value added of [organic], better economic return. I remember my conversations with several of the participants on a daily basis, and I remember one of them talking about his love and his passion for farming. He said to me, "You know, José, maybe I'm not making enough money to survive from farming at this stage, but I'm gaining other things. Because of this opportunity [to farm at the RDC] I was able to sort of go back." He meant going back to his roots, going back to his love for agriculture, to continue applying the learning from his own father and his family and his ancestors. So there were other important human aspects. The older participants brought [the experience of their grandparents] into their plots—how they farmed and how they produced and how they took care of the land.

Most participants were interested in the land, in experimenting and testing whether or not farming was feasible to them and a viable economic opportunity for their families. Most participants ended up going through the program and graduating, and then having access to some land: a half-acre to an acre the first year, then up to two acres the second year, and then up to four acres the third year. Then by the end of the third year, they would graduate. Some found land and leased land in the area: Hollister, San Juan Bautista and elsewhere; Monterey, South [Santa Cruz] County as well. The idea was to allow sufficient time for the participants in the program to hopefully establish some marketing connections and options so that when they graduate, they can continue with those relationships and those marketing opportunities.

We would have *loved* to see animals at the farm, cows or pigs or chickens at the farm, but the conditions just weren't there, mostly due to a lack of appropriate infrastructure. We had participants interested in those areas. In terms of crops, we had families who were interested in growing organic strawberries. Others were interested in growing a wide range of different varieties of cherry tomatoes, and others in growing zucchini and growing corn or potatoes.

We encouraged the participants to act as researchers. Research wasn't something that only people with [a] master's or Ph.D. could do but was something that the participants themselves could do as well. For example, they would grow six or eight tomato varieties in one row, in a hundred lineal feet. When those tomatoes were near ready for harvesting, we would call a couple of potential buyers, and we'd invite them to come and see the tomatoes and tell us which varieties they thought would be marketable. So that was obviously an incentive for the buyers. By 1998, the farm had become quite diversified with lots of different types of crops. We had thirty, at some point up to forty different crops growing—specialty crops: lettuces, and salad greens, green peas, cut flowers, fava beans, cilantro, strawberries, celery, corn, and many other crops. Most participants figured out that they had to diversify in order to enhance their marketing options. Initially, when I came into the program, most farmers were growing two crops: zucchini and fava beans. I remember one instance in which three or four of the program participants came to the same buyer one day with lots of zucchini. The buyer said, "Guess what. I can only buy a hundred boxes of zucchini." "But we have four hundred, we have four hundred and fifty boxes," the farmers said, "What are we going to do with the zucchinis?" And he said, "Well, I don't know. You're going to have to dump them or sell them or figure out a way, but I can only buy a hundred boxes." That was a major lesson. The following day they came back, and we had a meeting in the evening. They talked about that experience, and they looked at each other and said, "What are we doing to ourselves? We need to get organized and work collectively, plan things and increase our chances of competing and getting fair prices." So that was part of the reason why the farm became so diversified in terms of crops. By 1998, the RDC farm became a beautiful place. Participants nurtured the land using more ecological practices, and they had a small piece of land that in some way allowed them to continue fulfilling their devotion to a way of life.

By the end of 2000, I decided to step down from the RDC. By then I had developed other interests related to sustainable agriculture. One of the things that the RDC program participants kept talking about all the time, informally and formally, was that need and aspiration to both go back to their communities of origin in Mexico or Central America, wherever they came from, and to maintain strong social networks to their communities. I felt the same way. I learned that we wanted to maintain a physical and spiritual connection to our geographies, to our places. I learned that if we give up those connections, we're not who we are.

And so one day I remember asking the question in one of the meetings. I said, "Well, you know, I have been here for four or five years in this program, and almost every day I hear one or two participants talk about 'my community in Mexico or Central America. Oh, I wish I could go back, if I could.'" I asked the question, "Why don't we do something about it? What if we do something about it?" And they said, "Well, what would that be?" I said, "Well, we could start by exploring the potential for a binational exchange program. Maybe we invite farmers from your communities of origin in Mexico, help them process their visas, and invite them to come over and see and tell us what they're doing, what challenges they are facing, how do they see emigration from their perspective, what is it doing to their communities and to their lives. And on the other hand, we can also visit their communities in Mexico, visit with them and ask some of the same questions, share with them what our experience and life is like in the United States."

It was so intriguing and so unique, the response that they gave me to that question. Powerful. I expected them to say, "José, that's a great idea. Let's go for it. We'll do it! What do we need to do?" Instead they said, "Look, José, that's a good idea. But you know what? What's lacking is opportunities for young people to stay and thrive in their communities of origin in Mexico. We need to create *means* for them. For those who wish to stay, we need to create those means." One at a time kept saying, "Had I known of opportunities in my community to stay, I would have never left. I would have stayed."

I said okay. I identified a foundation that was interested in supporting the concept of my idea, and they said, "Talk to us about it, and we'll see how we can develop a proposal together." What we decided to do was a pilot program that we called *Proyecto de Arraigo*, or Arraigo Project. *Arraigo* stands for rootedness, developing roots in your community, staying and thriving in your community. So we contacted several community-based organizations, farmer associations, indigenous-based organizations throughout Mexico and asked them to tell us about who would be interested in participating in a pilot project. It was going to be a two-year pilot project that would engage key leaders from farming communities, rural communities throughout Mexico in an education, capacity-building process.

As a result of that, we received sixty applications. Because of the limited resources that we had, we ended up selecting about sixteen leaders from mostly indigenous-based communities throughout Mexico. The first task for the participants selected was to spend some time in their own communities and regions, assessing and identifying their top three priorities. They came back with that information, and we took that information to develop a highly responsive curriculum in response to that community-based assessment. For example, some of the communities said, "We have lost our ability to grow our own food in our own backyards. It's so basic, but we are interested in showing our young people how they can grow their own food." Another example: One community said, "We are interested in organic livestock management and production. We don't know how to do it. We don't understand what the process is for certification." And other communities said, "We're interested in learning how to build our own homes or houses using resources available to us locally." As a result of that, we set up a workshop based on the concept of ecological architecture, and each program participant developed a small prototype of the house they would build in their own community.

There were other wonderful examples. But the idea was that these participants were viewed and understood as multipliers of change. They would go back to their communities and share this knowledge. That was the main criterion, which is why we asked that each of the participants had to be sponsored by an association, by an organization.

Ultimately, the *Proyecto de Arraigo* lasted four years rather than two years. I was able to travel [to] all of these communities throughout Mexico and interview most of the people that applied to the program. It was inspirational to see. When I first visited some of these communities that applied to the *Proyecto de Arraigo* pilot program, it was stimulating, in that I remember seeing some indigenous communities actually were very happy with their way of life. You know, from the outside, people view them as poor, as not being on the train of progress. When I looked inside, at the heart of these communities, I realized, gosh, I wish I would have had this knowledge that they have before I emigrated. Visiting with indigenous communities was a transformational experience and also a profound process of discovery and realization.

My experience as an immigrant in this country influenced my interest in public policy. Over the first six to eight years of my life in my new country, the United States, I struggled with this question: Why did I emigrate? Why did I leave my community? I came to a point where I said, I need to do something to be able to deal with a sense of feeling endlessly overwhelmed. I need to do something about it, either go back to my community of origin for good, or do something here in my new community in the United States. I realized that what I needed was to engage in a learning, reflective process that would allow me to look deeper, [and take] a more objective, less emotional approach into the forces that impacted my life and that of thousands like me, immigrants with similar realities and conditions in Mexico and also similar conditions in the U.S.

I see myself maintaining as strong a connection as possible to the sustainable agriculture movement. Farmers—especially family farmers, small farmers—are struggling in every part of the world. I don't know exactly for sure where this is going to take me in terms of a future organization or professional role, but I do know that I will continue to maintain a connection and a commitment to sustainable agriculture. I have to do it in order to keep going. It's part of my life and my identity.

My parents still live in the small farming community, Providencia, where I grew up. My brother and sister also live in Mexico. I'm the only one who emigrated from my family. But I do go back a few times a year, a couple of times. I wish I could be there more often, but I have to now live and appreciate my new reality and the new opportunities this country has brought into my life because as an immigrant I've also been given opportunities to broaden my views and my perspectives in life, to meet people from all walks of life, and also contribute in different ways to sustainable agriculture. Working in sustainable agriculture here in the U.S., whether in an organization or simply as a responsible consumer, is also a way to honor the way of life I left behind.

María Inés Catalán

Photo: Rachel Glueck

María Inés Catalán and her family run Catalán Family Farms in Hollister, California. She is the first Latina migrant farm worker to own and operate a certified organic farm in California and the first Latina in the country to found a farm that distributes produce through a community supported agriculture (CSA) program. The Cataláns grow kale, chard, strawberries, tomatoes, corn, onions, pumpkins, chiles, and carrots, among other crops that they sell through Laughing Onion CSA and at farmers' markets around the Salinas, Monterey Bay, and San Francisco Bay areas. In addition to her farming, Catalán is also an activist who devotes herself to improving food security for low-income, Latino communities.[1]

I was born in Santa Teresa, Guerrero, Mexico, in 1962. My grandfather was a farmer. He grew peanuts, cotton, chile, and corn, and he raised cattle. I always think of my childhood as a very happy one, because I played in fields of cotton. It was all picked by hand. It was a traditional way of farming. It was rather organic, because we didn't use chemicals or pesticides at that time. It was all very natural.

My grandfather [was a migrant worker] for some years in the 1960s, for some seasons, in Texas. My mother emigrated to the United States thirty-five years ago, separate from my father. And out of all of my siblings, I was the last one to emigrate to the United

[1] This oral history was conducted in Spanish by Rebecca Thistlethwaite. Thistlethwaite and Catalán know each other from Thistlethwaite's work as program director for the Agriculture & Land-Based Training Association. The interview was transcribed and sent to Catalán for her edits and approval. Then it was translated into English. The Spanish and English language transcripts are both available at: http://library.ucsc.edu/reg-hist/cultiv/catalan

States. I came twenty-two years ago. That was only in order to be with my mother and my siblings, because I was basically alone with my children in Mexico then. I brought four children along. The youngest was two months old and the oldest was eight years old.

My father was [also] a migrant worker. What he tells me is that he only came to work, to earn a bit of money so he could support his crops [plant crops on his own land]. [Migrant work is] work that thousands of us people do, coming to labor in the fields. I worked maybe seven years packaging broccoli, cutting broccoli. It's a very repetitive job. You don't do anything but that one task for hours and hours and days and months and years. And you don't have any other option. You don't get to think about it; you just do what you're told. You're just another step in the agricultural process.

My mother worked many years in the fields, and she injured herself while working. She was invited to join ALBA's [Agriculture & Land-Based Training Association] PEPA [*Programa Educativo para Pequeños Agricultores*] program. My mother didn't want to come alone. She invited me and my brother-in-law and my sister along, to see what it was all about. I came just out of curiosity, to find out what organic farming was. So we all began taking classes, my brothers, my sister, my brother-in-law, and I. We heard the explanation of what it was about—that they teach you to handle tractors, that you make your own decisions with your own risks, that you could become a small-scale organic farmer—[and] all of this begins to motivate you to have a certain financial independence and to approach work much more reasonably. Little by little, my brothers had to return to their temporary work. I was the only one who finished. I graduated after three years.

I was one of the first women to graduate from PEPA. I knew that I needed to work and to be independent in order to teach my children how to work. I wanted to be able to support my family. The men would say to me, "What are you doing here? Go home to tend to your husband and take care of your children." They would tell me, "You're not going to be able to handle these tools. They are very heavy." And yes, yes, I could. There's nothing that's impossible. I learned to work with the tractor. I learned to prepare the soil. I learned about diseases, about pests. And many of the people who were laughing at me, or made me feel bad because they thought I wouldn't be strong enough, many of them got left along the way. They didn't continue growing. They didn't survive in agriculture. They just went back to their old jobs. And I'm still doing it.

My children weren't like other children from other families that got to go away for the weekend or got to go on vacations to Mexico. My children were always on the ranch, Saturdays, Sundays, and holidays. They would spend the entire summer on the ranch at ALBA. That's where my youngest children grew up. It's where they ran and played. My daughter, when she would get angry, she would tell me, "I'm fed up. I'm really tired of you and this ranch. I don't know why you had to go into training, why you didn't just go into training to work in an office. Here you're always doing the same thing, walking in the sun and in the mud." She would get very angry and throw tantrums. She would start to cry. But that's where my daughter grew up. There are pictures of her from when they first planted trees in that first little communal garden. There are pictures of my children there, from when they were very young. They know where we come from, how we started out. They've been with me through it all, through all of the progress, through the suffering and the limitations and the deprivations. Through it all.

All of them [help with the business.] My oldest son, he's the sales manager. He helps with the restaurants, and the shops, and the convenience stores. He's the general manager. He helps me with some markets, talking to the representatives, and with some licenses, permissions, certificates. He's the one that takes over when I can't do something. He likes sales. He goes through about four or five markets in a week. The youngest boy spends the week here at the ranch. He goes out on sales as well. And my daughter is studying. She's the only one who is in school right now. She's going to university in December in Fresno, and she's studying law. And she helps on holidays and on weekends, at the farmers' market. She helps me in sales, too. They've learned to be independent. That's the only way of life that they know. They've never worked for other people. They don't know how to work in the field for a company, with a boss that follows you around—you have only fifteen minutes for lunch, and they hurry you, and you always have to rush. They don't know that way of life. What they know is this: their own ranch, their own decisions, their own world, their own way of selling.

I think that we need to motivate the young generation and show them facts and proofs that this is our world; this is our agriculture. This is what we did before. It's a traditional form of farming. It's just that we've changed the name—it's organic farming. But in reality it's a style of farming that we're used to, or at least our parents, our grandparents. This is what they would do, you know? They didn't use pesticides back then, or any sort of chemicals. So it's the same thing. It's just that here, young people get the idea that if their parents worked in the fields, it's the worst thing they could do, that it's the worst thing that could happen to them, that those are the sorts of jobs that matter least. But the truth is, it's not like that. I think that we matter; we matter a lot. Working in the fields, or working on your own account, I think that we are very valuable. The work we do is very important, because not everyone can take ten hours of cutting broccoli, chopping lettuce, or picking chile. Not everyone has that strength.

I was the first Hispanic woman to start with the CSA [community supported agriculture] project in Salinas. Back then, my need to learn sales was so great that I started going to conferences. At that time, I wasn't eligible for grants. ALBA didn't have any funds. When I went to learn how to develop a CSA, everything was in English. I went to a conference in San Francisco and we had to sleep in the car. We'd have coffee and a donut all day, because we didn't qualify for the lunches. We were there for three days. And I would say, "All of these things that I'm going through have to add up to something someday, because this can't all be in vain." And sure enough, now we have our own CSA. We have it all.

I stayed up until one o'clock in the morning reading the CSA manual [given out at the conference in San Francisco]. I couldn't even tell you what it said. I tried to understand what it was saying just by looking at the shapes, looking at the drawings of the different vegetables that they had to offer and a few examples that they gave. The first year I almost went crazy planting. I was so happy that I was going to harvest and sell at CSA. I wanted to sell to my people, to Hispanic people, because they speak Spanish, but nobody from the Hispanic community wanted to buy vegetables. It's less desirable when we're used to being able to go and buy three bunches of cilantro for a dollar, or four bunches for a dollar. How are you going to pay what was at the time twelve dollars for twelve different things? And many people worked with broccoli, or worked with lettuce, or worked with radishes, with onions. So they didn't buy from me. There was a volunteer at Rural Development, and she was the one who began to promote me on the local radio station. The first people that bought from us were Americans. They were the ones who began to support me, and there are still about five families that have been with us since we started. Now they pay in advance for the whole year.

I started going to a farmers' market on South Main Street in downtown Salinas. We started to collaborate with the WIC [Special Supplemental Nutrition Program for Women, Infants, and Children, better known as the WIC Program], when there were no coupons. We had to show the system that there was a need for people to eat vegetables and fresh fruit. I had to come and set up a table over there, at the WIC office in Salinas, where the classes come in. With a small table there, I would sell thirty, forty, forty-five, fifty dollars worth sometimes, on a good day. All day I would be there to educate people and talk to them. And we're recognized as the first ones. We are still there.

Right now we are attending a farmers' market in the market-place in Ferry Plaza, in San Francisco, every Saturday. And I'm at the one in Berkeley three times a week. Those are the best markets in the whole country, in the whole country. So, just imagine. But like I say, I didn't get there with just my good looks, saying, "Hey look, I want to sell here and you're just going to let me." There were many people who supported me and helped me, and that's why we are where we are.

The people that can pay, do, and they pay well. But I also like giving things away to people who cannot pay at all. One example is Salinas. In Salinas, the majority of people are low-income. What we're doing is, we're printing out coupons. What I want is for people to come back and buy things and take the produce with them, because if it doesn't get sold it will just get fed to my animals. In Salinas, we're giving people seven dollars, in two coupons.

We, as Hispanics, shop where the prices are cheaper. That's why not many people go to the farmers' market. Why? Because if the town supermarket is selling tomatoes for ninety-nine cents a pound, sometimes seventy-nine cents a pound, why would people pay a dollar-fifty for it at WIC, when the supermarket has it every day, and they can go whenever they want? I tell them that I'm there, and people follow me and see me there and get closer. "Take three pounds for two dollars. Go on, take three pounds. It's cheap." Everyone wants to buy from me. That I know. If you're selling tomatoes—maybe heirloom tomatoes are a bit more expensive to produce—but Beefsteak tomatoes and Roma tomatoes, if you're selling them for seventy cents, at fifty cents a pound, you're making your profit. Your work is paid for.

It makes me feel good to support other people. Last year, the USDA recognized me for leadership in my community, for supporting the small-scale farmers. It's a national recognition that they only give to already incorporated organizations or various rural agencies of the USDA or employees, and the fact that they gave it to me, an independent small-scale organic farmer, left me like, "You're talking to me?" They recognized me on a national level.

Organic farming is something I grew up with. For me, it's nothing new. It reminds me of playing in the cotton fields, playing in mountains of peanuts, corn, watermelons, and melons, of herding cattle, and riding on horseback. I want my grandchildren to grow up the same way I did. It was something that is impossible for many people, because, you know, just imagine, you arrive here, and you have to live in a little room in an apartment. How can you dream of having your own ranch and having your grandchildren grow up the way you did? That's impossible for millions of people, because they work in fields where children aren't allowed to go because they are sprayed with chemicals and pesticides.

Now my grandchildren have young goats, they have chickens, they have horses, and cows. They are growing up the way I wanted to see them grow up. They run around all dirty, full of soil, of mud. They know all of the tomatoes, all the variety names, the colors, and all of the vegetables.

I lost my partner, the man who sent me to school, because he didn't support my development and growth as a woman and especially as a businesswoman. That person didn't support that internal change I was going through, and he had the option to leave. He left me. He abandoned me because I wasn't the same woman anymore. In our culture, it's very difficult for a woman to begin making the decisions. That's the biggest obstacle, culture. My [current] partner and I have only been together for five years. He's very traditional but he knows and he respects my decisions. He knows that I'm the one who is in control of the business.

My kitchen, and the dining room, is sometimes full of papers and documents. We're selling more than a half a million dollars a year. Sometimes I finish work on Sunday afternoon and, oh my God, I just don't want to have anything to do with it anymore. I don't want to hear anything about it. My daughter helps me to coordinate the girls [employees], but there are times when at four in the morning they are telling you, "No, Fulana is not coming to the market," you say, "What will I do? What can I do?" In places like San Francisco, or Berkeley, or Oakland, we need five people at each stand. People wait in line to buy from us. We're written about in the newspaper: the best garbanzos, the best tomatoes, the best strawberries. We've been in first place, the best strawberries in Berkeley, for five years. I can't complain. I don't have any money, but you see, it's a really nice feeling, being recognized that way.

[To be a successful farmer] you have to have a need, a desire, perseverance, strength, and insanity. I think I've always said that my need was the cause of my achievement. I had a need to provide food for my children, to make a bit of money. I had that need to be able to survive and that need brought me to where we are now, through my work. It took a lot of work and sacrifice and tears. I had to stop being a daughter, stop being a sister, stop being a woman, stop being a mother, many times, because I have always prioritized my business. As long as I am living, I'll be doing what I love to do.

I would like to give motivational workshops someday to young farmers, to support them, to guide them, to advise them. We, as simple people, without an education, many times we have dreams and our reality kills our dreams and doesn't let us get ahead. So, find a way, if you have a dream, to not let your reality kill your dream. Believe in yourself, and find a way to make your dream come true. Grab hold of that dream.

Florentino Collazo and María Luz Reyes

Photo: Sarah Rabkin

María Luz Reyes and her husband, Florentino Collazo, run La Milpa Organic Farm on land they lease from the Agriculture & Land Based Training Association (ALBA) near Salinas, California. They are both graduates of ALBA's six-month course known as the Programa Educativo para Pequeños Agricultores, *or PEPA. After graduation, Collazo worked for eight years as the field educator/farm manager for ALBA, and Reyes continued to farm on land she leased from ALBA. Then Collazo left ALBA to farm full time with Reyes. They have run La Milpa Organic Farm for the past six years and are certified organic by California Certified Organic Farmers. On their farm—named La Milpa in tribute to traditional Mesoamerican methods of growing diverse crops closely together—Reyes and Collazo cultivate over thirty crops that they sell at farmers' markets throughout the Central Coast.[1]*

Collazo: I was born in 1963 on the Providencia ranch, in Municipio Purísima del Rincón, in the state of Guanajuato, in Mexico. My father did various jobs. He was a politician, a farmer, a rancher, a business-man; he worked in almost everything. My mother was a housewife, a very hardworking woman.

Ever since I was a boy, we were very concerned with produce, with vegetables. They were traditional crops, basic crops, like corn, sorghum,

---

[1] This oral history was conducted in Spanish at La Milpa Farm by Rebecca Thistlethwaite. Thistlethwaite, Collazo, and Reyes know each other from Thistlethwaite's work as program director for the Agriculture & Land-Based Training Association. The interview was transcribed and sent to Collazo and Reyes for their edits and approval. Then it was translated into English. The Spanish and English language transcripts are both available at:
http://library.ucsc.edu/reg-hist/cultiv/reyescollazo

beans, pumpkins, tomatoes, traditional crops of Mexico. When we were little, we would steal water from my mother. Water was scarce and we would steal it to plant cilantro and other little things. There were fourteen of us children. I'm the sixth, but with the ones that passed away, I'm like the eighth, because five of them died. So now, the ones who are still living, there are fourteen of us, and I'm the sixth.

I just have an associate degree. I finished all of the coursework [in Mexico] to be an agro-engineer. I have never returned, but I hope to obtain my title someday. I only have to present a thesis to obtain the title, but I finished all of the courses. When I got out I came here and I forgot a bit about my education and instead focused on the practice. Now I feel that I have much more practice than I have education. I've been working in the field since I arrived here in 1995. We're already talking about fourteen years that I've been observing, acting, improving. Nobody can take that away from me, and nobody can challenge it.

When I graduated from college [in 1995], we were going through a period like the one the United States is going through now. After an earthquake in Mexico City, the economy was devastated, but besides that there was a crisis going on and there weren't many paying jobs. At that time, there was a lot of work available here in the United States. That was when they passed the Amnesty Law of 1986. That's how we had the opportunity to become legalized and come to work in the United States, even though the work that we came to do wasn't quite the work we knew how to do, but in any case, it was related to agriculture.

I worked for twelve years harvesting and packaging lettuce in the Salinas Valley and in Yuma and in the Imperial Valley. Picking lettuce is one of the most difficult jobs, because you have to bend over to harvest and package the lettuce. At that time, there wasn't yet any packaging like you have now for salad. It was all done on little squares of soil, and it was paid by contract. It wasn't by the hour. It paid pretty well, or at least for the time. Part of the difficulty in the fields has been that you have to battle against the weather conditions. Sometimes it rains a lot, it gets very muddy, and the work is always very hard, very heavy. Some years I would work straight through. I would move on to the Imperial Valley and to Yuma, or to the San Joaquin Valley in Huron, California, near Coalinga in Fresno County. But sometimes the season here would only last six months, here in the Salinas Valley. Mostly I could stay here with the family, but at the beginning of the season we would start up again.

My dream was always to make a living in agriculture. Since school I've always been keen on organic chemistry and biology, botanical ecology. There was a program called "The Best of Our Own" here in the valley, here in Salinas, on the local television network, and one time the director and the manager came on television. I already knew about [the Rural Development Center], but I had never been inside. So when I saw them, I came personally and I met with the man who was the manager. I began to chat with him, to see how it was, how all of the farmers were, and about the meetings they have here. He invited me to attend the educational course for young farmers that was about to begin. I came in with a lot of motivation, and because ever since Mexico I've known how to use tractors and all that sort of equipment, I asked him if I could be a volunteer, because they had a lot of area to cover. They had an old but very large tractor.

That first time that we started coming, the first year, it was forty of us in two shifts. We came as twenty and twenty. Twenty went on Saturdays and twenty of us came on Wednesdays. I would come both days. I was so motivated that I would go Fridays and Saturdays. I really liked it, because I thought that it was what I was looking for. I was always looking for a small plot of land. In most of the Salinas Valley, they don't rent out small plots of land, only one hundred-acre parcels with large, high-volume agricultural wells. So when I found ALBA, I knew that was what I needed, because I saw that they had small shared plots, but it also was a bit problematic because sharing water can be a problem.

I worked for ALBA for eight whole years. I already knew how to do some things, and I was learning other things from the other farmers, because you never stop learning here. I passed on many things to many people here who were already farmers on their own. [I] was trying to bring the best out of the program, to always be looking out for a better way to distribute the water, a better way to coexist, even though it is very difficult here to keep good relations up between farmers, because we all come from different places. Mexico is divided into regions, and it is very difficult to coexist with people from other regions, because in certain areas their way of speaking is sometimes very different. And it's not just their way of expressing themselves; they are all quite blunt, or we are blunt, and sometimes people's feelings get hurt. So to be here in this position is not easy, because you're on the front line where all of the conflicts are brewing. It's a question of compromise, with the maintenance of natural resources, because it's a very important part of sustainability and organic farming. And, also, the focus was on education. Part of my responsibility was theoretical training, and later practical training, for going into the fields. I really liked giving the PEPA classes, teaching the classes on irrigation, on the handling of the equipment, the things in which I had a lot of experience. Those were my favorite subjects. I've seen people that don't know how to handle a garden hoe. There are techniques for using tools. You have to know how to move it and everything; you have to know how to use it around the crops. My favorite subjects were the equipment, irrigation, anything related to the preparation of the earth.

Some [of my former students now] have their own ranch; others are caretakers. Many secured better positions, because they could tell the boss that they already knew how to handle a tractor, that they already knew how to create an irrigation system. They're not planting anymore, but it did help them in their education, in their work.

You have to be authentic and original, so you call attention to yourself. When I came here, to the Rural Development Center, everybody asked us what a *milpa* was. And nobody knew. Nobody knew the answer. Everybody said, "Ah, but it's obvious. It's the corn." The man who was the director here at the time, another agronomist in Mexico, Mr. [José] Montenegro, he was the one that explained to us what a *milpa* was. It is a word with Mayan origins, and it's the combination of three or more crops, which is exactly what we are doing, more than three crops. We have as many as twenty-five. Biodiversity. So, the concept is well said and well done. It's a very Mexican word, but after the Green Revolution, after the Second World War, when people started using agrochemicals, and herbicides, and started to eliminate the wider plants to leave only the pure corn [it began to be used to refer to] corn alone. Before, *milpa* meant a combination of corn, beans, squash, and tomatoes. It was sustainable agriculture; it was a type of agriculture through which the corn could feed the people. Then there were the stalks, or rather, the dry parts of the corn, and the beans gave out nitrogen for the corn. Also, it was the staple food of the Mexican people, the squash. We could sell its seeds by the kilo, and give the rest in pieces to the animals. Or we would cook it in the oven or however, or mix it with milk, and it would be very nutritious. Later would come the tomatoes, and we could make sauces. But the second function of the squash would be to completely cover and cut off the weeds. It was a very ancestral way of farming without the use of much technology. To me, it felt very mystical, very traditional, and very original to name our farm La Milpa. Most of the clients are English-speakers, and I try to explain it to them with what English I know. They think that because it is named La Milpa, we are selling corn. They ask, "Where is all the corn?"

There is a lot of variety with our crops. For example, right now we have about fifteen varieties of tomato like heirloom, regular, yellow cherry, orange cherry. I have seven varieties of squash, two varieties of cucumber, two varieties of beet, cilantro, two varieties of onion, rainbow chard, celery, four varieties of chili, fennel, and purple artichoke, romaine, strawberries, raspberries, blueberries, artichokes, red lettuce, cauliflower, broccoli, romaine, celery, strawberries, raspberries. And later still there are amarillas, golden berries. There are green peppers, those green chilis, and corn, and flower beds for the corn, and onions, fennel. And later on come the cucumbers, the squash, basil. Basil and carrots, and later there are tomatoes, chards, green beans, yellow beans.

I enjoy it all, from the time I prepare the soil to when I see how high the plants have grown. I feel like I'm dragging a dead farmer's soul around in my body that just won't let me be, that won't leave me alone. That's genetic. One of my grandfathers, my mother's father, was a perfectionist, a very hard worker when it came to the fields. My mother has the same capacity, that same drive to work the fields thoroughly. My mother knows how to work the fields better than she knows how to cook a dish, how to run a kitchen. She handles the soil better than anything; she has her plants, she has everything. So I think that is hereditary, that passion for the fields.

Look at my hands. That is why I love to work the land. I don't like using gloves, because it's like taking a shower with an umbrella, you understand, putting an umbrella over yourself when you wash. The water just falls around you. When I want to work, I want to feel the earth. When I pull the weeds, I want to feel my fingers penetrating the soil, feel that I'm pulling them up, that I'm doing it myself. My hands and my mind are linked. I really love to look around, walk up and down observing, surveying it all and saying, "Wow." That's what fulfills me. When I'm at the farmers' market, when people are arriving, reaching for the produce, and then later passing by, my self-esteem really rises. I feel very proud. All of your efforts are reflected in that moment. You've really accomplished something. When they tell you, "These are the best strawberries I've ever tasted. I'm going to take them," they flatter you. Ah, it makes you feel a light in your soul.

Reyes: My name is María Luz and I was born on a ranch in Mexico, in the state of Jalisco in 1965. My father was a cattle farmer. His main task was milking. My mother was also a housewife. From a very young age my father would drive a team of oxen and I would tie some bottles around my neck with string. On one side I'd keep seeds for the beans, and on the other seeds for the corn, and I would go down the line planting them. I could feel the breath of the oxen behind me, stomping at the soil to get ahead. I liked the dairy farming. I liked handling the cattle. We were three girls and seven boys [besides me]. I was one of the youngest of the girls. I had already finished school by the time I was sixteen years old.

[When Florentino and I came to the United States] I packed asparagus for a very short time. I was working in an asparagus plant. I was separating the asparagus, because it can be big, small, medium or smaller. They keep changing the amount of asparagus. First they put you on the first belt, which is for the larger asparagus, and then they keep moving you and moving you up, and on the last belts it's very fast, and you have to get it all. It's very rough. And I didn't last very long, because I became pregnant. I started getting dizzy and I quit. Florentino needed me, too. When Florentino joined as an agriculturist at ALBA, I began to help him. I would go asking ALBA for help—to take my produce, because they had very good connections on the land. I would sell them my parsley, because no one would buy it from me. And they would go looking for sellers for me, to buy up my produce.

Both of us are farmers, but each of us focuses on one thing or another. I am the one that manages the planting process, the irrigation, preparing the soil, and handling the money during sales. Or sometimes I manage whatever quantity of produce we are going to bring to the market; I direct the packaging and so on. I take care of the organic matter. But he takes care of the payments for fertilizers. For the farmers' markets I harvest everything, and he takes care of the strawberries. We split it up. You go and do that, and I'll go and do that.

I love seeing when the plants are all cleared, when there are no weeds, and they are almost ready to just cut themselves. That's what I like. They look beautiful that way. People pass by and say, "Oh, your parsley looks so nice." Florentino really knows how to work the land. Many of us know how to grow things, but really working the land is not an easy thing to do, making even plant beds and cultivating and all of that. Not everyone knows how to do it.

My son who is at Berkeley, he said, "Ma, you're not a human being. You're a working machine. What have you gotten out of it? You work too much and for too little." Sometimes we don't have the chance to eat, or anything. You work a lot, and the field is very tough. We don't get even one day of rest. But we feel satisfied, because little by little we are getting out of some debts, and well, really, we are making some money [laughs] which is what everybody wants. We have confidence in ourselves, because we are the ones who are watching over the whole process, the ones harvesting. It's not that we are sending someone else out to harvest without knowing what's going on. No, I'm there seeing what's going on, what I do and don't like. Every day we are learning more and more. One never stops learning.

We're starting on something a bit different. We want to separate ourselves a bit more from ALBA and devote ourselves almost entirely to selling our own produce, because we're seeing that that works out much better. When I have a bit more from that end, I do want them to help me out. But I also don't want to force them to sell anything for me. I understand that ALBA is an incubator, that it serves to teach farmers that it is their obligation to become independent and learn.

The most difficult thing is the wintertime. In the wintertime, it's pretty hard for us, because there is not much to harvest. Especially in the farmers' markets, there is not much produce. But in the end it's worth it. In November, when we still have tomatoes, they are sweet, very sweet. Yesterday, when they had that festival and all of these people came out to buy, a man said to me, "I've never touched the sky, but with these strawberries I just did."

J.P. Perez

Photo: Gerry McIntyre

Juan Pablo "J.P." Perez founded his J & P Organics Community Supported Agriculture program in 2006 while he was a college student, with a subscriber base of five friends and advertisements on Craigslist and a campus electronic marketplace. While he grew up working on a farm with his father, J.P. first learned organic farming in the Agriculture and Land-Based Training Association's Programa Educativa para Pequeños Agricultores, which trains small farmers in organic production methods and marketing techniques. Today, Perez employs his parents and siblings in his expanding farm enterprise, which serves about three hundred (and growing) community supported agriculture (CSA) subscribers. In his mid-twenties, Perez is one of the youngest organic farmers to run such a burgeoning enterprise.

Back in '97, '98, '99, my dad used to lease five acres of land. We were growing raspberries and cut flowers in Watsonville. My dad used to bring me to work after school, on summer break, on weekends. It was hard work being out there your whole summer. One day my dad said, "J. P., what do you think? Do you want to come work with me and be a farmer like me, or would you like to go to school?" I said, "Well, I'll go to school," because farming is pretty hard. I wanted a better-paying job. So I went to Pajaro Middle School and Watsonville High, graduated from North Monterey County High School in Castroville.

I went to California State University, Monterey Bay. I tried liberal studies, for teachers. It wasn't for me. I tried business. It wasn't for me. I tried computer science. Computers just drove me crazy. And then I tried an earth systems science and policy course, and I really liked it.

I liked the science. Well, why not take science? I wanted something easy [chuckles], but I ended up taking the hardest major there. I'm a hands-on person. I like to be out there in the field and learn from it, instead of being in a classroom and just writing it down with a pen. I'm a visual and hands-on learner. I stuck with it. And then I found out about the internship program the science department had there.

And that is where I found out about ALBA, the Agricultural and Land-Based Training Association. I did two internships with them. One was in the summer of 2004. I was helping the farmers there, doing some research on organic fertilizers, pest controls, forming a little library, helping them out in their fields, learning from them on organic production. My second internship was helping Patrick [Troy] teach the PEPA [*Programa Educativo para Pequeños Agricultores*] class, and helping out the new farmers, and learning more about organic farming and what regulations, permits, certificates and all of that you need. I liked the program.

Back when we had our farm, a problem came up. The pump [for] the water well broke down, and we lost a lot. And that was the third time the pump broke. It was a lot of money to be fixing it, so we sold everything, all the equipment. My dad didn't give up. He just was just like, "You know what? Let's take a break. I don't want to have to go through this again." So we just left the farm. My dad got a regular day job; my mom did too. They worked for Monterey Bay Nursery in Watsonville.

But now we're going back [to] what we were doing before. Because [PEPA] was a Spanish course, I presented the opportunity to my dad, "Dad, I know you like farming. What do you think of coming and taking that class and doing it organically?" He was, like, "Sure, I'll do it with you. Let's do it together." And we did. We went through the whole course, six months.

[I also did an internship with] Serendipity Farms. Serendipity was more about marketing—wholesaling and packaging, CSA interaction with customer service people, farmers' markets, how to set up a booth, put a banner, have your certificate, make everything look really nice so you can attract people into the booth, have samples. "Here, you want to try a strawberry? You want to try a raspberry?" They get hooked. They like it, you know? Those little key points helped me take from one step to the other.

So I got the whole idea, the whole system—how a farm sells and everything works—before I started running our own business. Certification. CCOF [California Certified Organic Farmers]. You have to play by the rules and by the book. What type of materials to use. What type of seeds. You have to use a cover crop, because it gives nutrients to the soil. Crop rotation. Use flowers to bring in beneficial insects. What not to plant in the summer but plant in the winter, because there's less pests in the winter. I was kind of smart on taking the internships, learning from them, and then using those ideas I learned from the internships into my own little operation.

At the old farm we did it conventionally. I was probably, like, thirteen years old, so I didn't know anything about organic farming. My dad used to use pesticides. Not a lot, but he used to spray Round-Up for the weeds and use conventional fertilizers. We didn't have any employees. It was just basically a small family operation. And when my dad used to spray, he didn't take us to the farm. We stayed home. Or he just told us, "Stay away from here. Don't come in here, because I just applied something."

Through my courses at California State University, Monterey Bay, I was reading about water law, ethics, about all these chemicals going into conventional farming, draining out into the ditches and then into the Monterey Bay, killing birds, killing fish. People are being affected by the pesticides the companies use in their crops. Women who are pregnant, they get a little bit of the effects. And it was like, Let's try something more natural, more healthy, that people are going to enjoy and we're not going to put our health at risk.

At first my dad didn't know about organic. "What's organic?" "It's more natural." "But isn't it more problems? And how about the pests? How are we going to control the pests?" Well, we started reading, doing more research. ALBA offered pest advisers that came into our farm, and they looked at what kind of problem we had in our product. They said, "Oh, use this, and use this, and this will help." Okay, we got advice. Now he likes it. He's all into it.

"J" is for me, Juan, and "P" is for my dad, Pablo. We started with an acre. We each got half, because usually for each graduate they offer you a half-acre. We were used to growing just flowers and raspberries. It was kind of funny because here we were growing carrots; we were growing strawberries. We were growing cucumbers, zucchini, beets, chard—as an experiment, basil. It was only one acre, about twenty or thirty rows. So we tried to do it as diverse as we could, a row of each type of thing, because when we started, we didn't have any marketing.

That was one of the problems we had, to start up, where are we going to sell our product? We didn't have any farmers' markets. We didn't have any CSAs. ALBA was buying only a little bit of our stuff. So that's where I decided to start my own CSA. I started with friends in college, because I was still in college in Monterey when I started the business. Now, we have the wholesaling; we have the CSAs; we've got the farmers' market; we've got the cut flowers. We've got the whole system going on.

I started with five friends, and I put an advertisement on Craigslist [laughs]. I was putting advertising on the school's website, a marketplace to sell your books and stuff like that. I put a posting. "If you want to have veggie boxes, if you're interested, please give us a call." And then from there—boom, boom, boom, we started growing.

They liked the box. It was really cool, really fresh, really good tasting. The best part of it is, I delivered to them. At this point, I don't have any pickup spots. We do home deliveries to all our customers. A lot of CSAs have spots where people come and pick up their boxes. We try to do it a little bit different from them. We do home deliveries.

Service is what really counts a lot. We have customers that can't drive. They don't have a car. Or they're so busy at work that they don't have time to go to the farmers' market or to the grocery store. When they come home, they see the box on the doorstep. They can choose to have it weekly, every other week, or even once a month. No contracts, no obligations. Give it a try for one week. If you like it, keep going. If not, we can stop there. Pay as you go. [laughs] Right now we have a total of three hundred [shares] and counting. We started the business in 2006. This is our fourth year.

We have strawberries. We have raspberries. We have cucumbers. We have about ten different types of zucchini. We have bell peppers. We have the heirloom tomatoes, the cherry tomatoes, the grape tomatoes. We have cilantro. We have basil. We have herbs—epazote. It's a Mexican herb that Aztecs used to drink in a tea and it's really good for you. A lot of people put them in their beans to take the gas out of the beans. You just drop the little leaves in there, and it's really good. We have carrots. Did I mention beets? We're growing pumpkins, too, for Halloween. Artichokes. Lettuce, broccoli, green beans, peas. Chard, rainbow chard, kale. We have a lot of cut flowers. We have sunflowers, statice, dahlias, Sweet Williams, strawflower, cornflower. Pompom, which is a really nice flower. Daisies. Native yarrow.

We have a lot! We grow about twenty to thirty different crops during the year. We have crops growing all year round. Our members, they love that. We have stuff growing for the spring, summer, and winter, so they taste that cycle.

I have customers that are allergic to some of the stuff, or they don't know how to prepare it, or they don't like it. Like Brussels sprouts. Some people hate them; some people love them. I have a customer who's allergic to onions, so we have to take them out. They let me know in the e-mail when I let them know what's coming in the veggie box, "J. P., can you please substitute the onions for more strawberries?" or more fruit or more kale, more spinach. We totally do that for them. Make them happy. If they're happy, we're happy.

Our product is really affordable to anybody. We prefer to sell it instead of [the vegetables] going bad. We want the people to have access to it. Our boxes are twenty-three dollars a box. The delivery is included. If they do the math, they're saving money. They're saving their time. They're saving their gas. Of course, conventional is cheaper. But you're not going to have the same product.

We do mostly all Monterey County areas. We do Carmel Valley, Pebble Beach, Pacific Grove, Monterey, Seaside, Ryan Ranch, CSUMB [California State University, Monterey Bay], Fort Ord, Marina. The Monterey Bay Aquarium. Watsonville, Santa Cruz, Las Lomas, Prunedale. We deliver to the Dominican Hospital in Santa Cruz. Soquel. Aptos. [In Santa Cruz] we do Mission Street, we do by the Boardwalk. San Jose State University, Stanford University. The farthest we go is Oakland, because we have a farmers' market there on Saturdays, so customers or members order a box. It's already packed, prepared, and everything. They just come to the market, they pick it up, and they pay there. And we're killing, like, two birds with one stone there. We have the booth, and we have the boxes there for them.

I'm the guy who's doing all the deliveries. I deliver to three hundred CSA members, but some want every week; others, every other week; and some, once a month. So by the week, it depends how many deliveries I do. Roughly, I'm delivering 150 to 160, 170 boxes a week to the doors. Most of my time is just driving, delivering to farmers' markets, paperwork at night. We might need to get another driver because it's getting to the point that the daylight—it's not enough for me, and I'm getting there really late, and like, "Oh, sorry, I just had a lot of deliveries," and I don't like that. I don't want to be there too late.

Basically, a typical day of mine will be getting up in the morning maybe around six or seven, getting all the boxes ready, getting all the produce, bringing it back, packing the boxes, delivering in the afternoon, be done around, if I'm lucky, six, seven or eight, get home, eat dinner, go back to my computer to look for the orders that come in, e-mails that come in, and then do the same thing every day. It's a long day. Sometimes [I go to bed at] one, sometimes two. It's worked, and I'm proud of it. I like it. It's my job. And it's paying the bills at this point.

ALBA told me they wanted to start a farmers' market up there in Oakland. I said, "I'll do it." I talked to Jason, the manager there. I turned in all my paperwork, and they said, "Okay, you're ready to go." It's on MacArthur Street. It's called Oakland Food Connection. It's only a small little market. They're just starting it up. We started going on October of 2008, last year. And it's all year round, rain or shine.

There are other folks selling there. They have a vendor who sells biscottis. They're really good! I'm hooked on them. There's another farmer who has honey. There's another who has flowers. And there's the Purple Lawn Café. They buy our product— like the kale, the carrots, the zucchini and onions—and they do a nice plate. It was really good this past Saturday. They had beans, they got rice; they had bell peppers, zucchini, kale sauté and salad.

We have a total of three farmers' markets. We have one in Greenfield we just started. We had our veggies, and now we incorporated more flowers into our booth, so we have the flower section and we have the produce section. It's doing good. Mostly Latino people. We have the one in Oakland. And we're going to start another one tomorrow in Oakland, too. That's called Children's Hospital and Research Center. It's brand new. They just got all the permits in and everything, and we're starting tomorrow, actually.

We sell to ALBA Organics. [Say] we have a lot of raspberries. First we take care of our customers, and whatever is left, they go to them. So nothing goes bad. They take one box up to thirty, forty, fifty boxes. Or if they get an order in, "Can you bring me two boxes of chard?" "Can you bring me fifteen boxes of carrots?" They know I'm the only farmer at ALBA that has the CSA. I say, "First thing, let me take care of my customers, and then whatever is left, you get." "Okay. Good."

I was selling to stores, but I stopped because I didn't have time to deliver to them, because I'm doing all the CSA. I could probably sell to them and drop it off on my way to them. But the wholesaling is doing good, basically.

We have a customer in Oakland that buys artichokes, strawberries, potatoes, what we have in our farmers' market stand. Whatever we don't sell, he takes. It's an Italian restaurant. He's a good chef and a really good person. And we've been getting calls from [other] restaurants. But I need another helper. Maybe in a couple of years I want to get more vans, hire more people and make them go to the farmers' market, and maybe have fifteen markets or so. We're expanding as we can. We don't want to take a big step. Taking baby steps as we grow, as customers want us to grow.

We can lease the tractors from ALBA. That's a big help. The rent, I think, is pretty reasonable. It's cheaper than the real market out there. We know what the real price is out there for land. It's really expensive. That's one way they've been helping us. We're leasing from them. The water bill, I think it's really cheap, too.

ALBA has put me out there. They helped us to be in the CAFF [Community Alliance with Family Farmers] "Buy Fresh Buy Local" guide. That connection with CAFF, the connection with [California] FarmLink, their connection with Cal[ifornia] Coastal Rural Development [Corporation], their small lending business. They can lend you money. FarmLink, they can help me find the farmland to buy or to lease. And schools that come by. That's how we got Stanford. A group of students came to the farm, and they talked to my dad and we told them, "Oh, we have a CSA." It was, like, "Oh, we want to start a CSA there, in our school." "Sure." And then they talked to me. So I think ALBA has really brought us out to the public.

FarmLink has an IDA [individual development account] program. It's a three-to-one match program. This is my second year. For two years, I had to put a hundred dollars a month, and then at the end of the two years, they gave me a check. It was like a saving plan, a check to buy a hard asset like a tractor. I'm going to use that money to buy another van so I can keep it going. So FarmLink has helped us a lot. Reggie Knox [of FarmLink] has been going with us and checking out [possible] farms [to buy], and he's been asking questions and this and that. We want to get good advice from him, because he's a person who knows a lot about spaces and farms.

California Coastal Rural Development Corporation, they're a really good small lending company. Last year, they lent us $38,000 to buy equipment. We bought two tractors. We bought another truck for the produce. We bought material, equipment for the farm, just for the farm. And also I got a computer, a printer and all that stuff, just for the business, to keep it going.

We have a total of five acres now. My mom and one of my uncles are helping us at this point. My sister helps out, too, doing the veggie boxes when we have to pack them. And my brother, he has a regular day job, but he's a mechanic. He graduated from UTI [Universal Technical Institute], so he's going to be the one fixing the tractors and the trucks. My dad is the grower and he waters, he fertilizes and everything. He does the weeding. My mom is the packer. She harvests and makes sure everything is really good. I'm the salesman, doing the CSA, farmers' markets, and promoting J & P Organics. My sister is mostly helping us do the veggie boxes and also doing a little bit of the paperwork. So we have the whole circle, connecting, a total family operation. All the income, all the capital is staying with us. We have to pay the mortgage. We have to pay our bills. We pay our salary. So it's good because it's staying with us.

If [people] want to visit our farm on a weekday, they can come. Any day of the week. As long as we're out there, they can come and harvest the berries, whatever they want. Because my parents are out there. They do all the interactions. But usually I get the phone calls, and I let my parents [know], "Oh, Dad, there's somebody coming down there. They want to get some flowers," or, "They want to get some berries." And it's, like, "Okay, I'll be here." Families come to the farm and they bring their kids, and they want to teach them where stuff grows, how it grows, out of the dirt. Get a shovel and dig all the potatoes and pick them out. They like that. They like getting dirty.

It's just a lot of work. You have to be on top of everything. Doing all the paperwork. That's the other thing that drives me crazy, the bookkeeping, keeping track of all the receipts, all the sales, all the numbers, paying the bills, the car payment, the van payment, paying the cell phone because that's basically how I communicate, paying the Internet because without an Internet I don't have a business. Making sure we have fertilizer. Make sure I buy the seeds. If a crop is going on, make sure to have the other one on top of it. By the time this one is done and we disk it, we have the other plant ready to plant on that same day.

I'm doing all the seeds, looking for new companies, looking for the best prices out there, talking to customers. My dad helps out, too. He's doing the flowers, because that would be another workload on my plate. But I'm doing all the calls, talking to or e-mailing CSA members, trying to get into restaurants. You know, just doing phone and Internet and e-mail. Make sure we get all the orders in, make sure to harvest forty, fifty bunches of carrots for our boxes for tomorrow.

It's funny because I went to school to get a better-paid job, and it brought me back into farming again, but now doing it as your own boss, not working for somebody else. You're the boss now. You make decisions now. That's what we like about it. We're not working for anybody. We're working for ourselves. I don't know if it's destiny or it runs in my veins or something, but it brought me back into farming. But now that I'm doing it, I like it, and I'm happy.

By the fifth or sixth or seventh year, I think [ALBA] starts pushing you [to] move on and look for your own farm. Those are the plans for us for next year. Hopefully, we can apply, or start looking for a small farm, because my goal is to have the whole operation—have my cooler, have my office, have chickens for the eggs, cows for the milk, pigs for the meat, orchard trees. I want apples, pears, nectarines, plums, cherries, apricots—the whole thing. J & P Organics and Small Farm, where we can have the farm open to the public, and they can come by, pick their strawberries or raspberries, whatever we have available there, so we can have more stuff to offer to our members.

Congressman Sam Farr

Photo: Tana Butler

United States Congressmember Sam Farr is one of the political heroes of the sustainable agriculture movement. Before his election to the House of Representatives in 1993, Farr served for twelve and a half years in the California State Assembly. In 1990, Farr authored the California Organic Foods Act, which established standards for organic food production and sales in California. This piece of legislation became one of the models for the National Organic Program's federal organic standards. Farr now serves as cochair of the National Organic Caucus in the House of Representatives and works with organic policy activists to increase support for organic farming research in the federal Farm Bill and other federal programs.

The south side of our house in Carmel was the sunny side, and my mother would plant gardens. I would love to get in the garden and plant. I got interested in flowers and vegetables. I always say my first experience in being a businessperson was selling flowers in front of my house on a little lemonade stand, because I learned a lot about psychology. Actually, not all the flowers came from my yard; a lot of those flowers came from the neighbors' yards. And when I sold the flowers, I noticed that the neighbors bought the flowers that came from their own yards, so I learned a little bit about consumer choices in this little flower stand that I had.

I was an early environmentalist because I was dyslexic. I wasn't comfortable in grammar school. I thought I wasn't very bright, because I couldn't read very well or pronounce words right, or spell words correctly, and so I would cut school anytime I could avoid it. We were living in a town where you didn't go home, you just went to wherever kids were having fun. You'd grab a bunch of neighborhood kids and go to the beach, and go to the tidepools, and go to the lagoons, and hike whenever you can. It was before television, before anything that would keep you inside the house, so you just lived outdoors all the time. And in the process, you become very aware that there are different kinds of birds out there, and different kinds of songs the birds make, and there're marine mammals and marine animals. You got interested in different kinds of trees and plants.

I was told by a sixth-grade teacher in science class that I was a pretty good student. I hadn't been ever complimented in school, so with that compliment, I ended up deciding, I'm going to keep studying these things I'm being complimented on. I'll follow the praise. I was a biology major all through high school and in college. I took botany courses. I worked in the Forest Service when I was in college, as a firefighter. You learn a lot about riparian vegetation and forest health and things like that, not at a degree level, but just for practicality purposes. Now my hobby is landscaping my property in Big Sur and figuring out what kind of plants could grow there. I'd love to get some stone-fruit plants, but we live right on the coast and you don't get a lot of very cold winters. You have to find trees that have a very low chill factor in order to be productive. I've experimented with different varieties and different types of plants. What's really taken off is avocados.

I was in the Peace Corps from 1964 to '66 in Latin America.[1] I wasn't really in agriculture, but living as a minority in another land, you learn a lot about the fresh fruits and vegetables. The difference is that you've never seen any of these things before. I laugh about this, because I remember when looking at stuff I didn't know, it would be like somebody looking at a banana and not knowing how to eat it. Do you just eat the thing raw? No, you got to peel it. Or do you have to cook it? A lot of the fruits in Latin America are like that. You've never seen these fruits anywhere. Do you peel them? Do you cook them? What do you eat? It's interesting how much of our foods are cultural. This way we got to know our next-door neighbors, because we figured out we didn't know what the hell we're doing. So we hired them to cook for us, and then they showed us. We said, "We'd like to learn."

[1] Farr served in the Peace Corps in Colombia, near Medellin.

I was an urban community development volunteer. I went into the Peace Corps right after college to organize people living in the barrios on the outside of town. The rich people lived inside, and the poor people lived on the outside, just the opposite of what's happening in American cities. The job was really not very well defined, but it became one of teaching them how to set priorities and then go down and petition their local government to give attention to building sewers and schools and health care centers and playgrounds and things like that.

I came home in 1966. President [Lyndon B.] Johnson had declared the War on Poverty in America, and people were discovering—oh, my God, we've got poor people in America? I didn't know what I wanted to do, and all of a sudden I was getting these job offers from all over the United States because I had been in the Peace Corps; I had done this cross-cultural [work], and they knew that I'd been in community development. They were hiring community development directors all over the country. I think I had twenty-seven job offers without ever asking or seeking them. I didn't take any of them. I went to law school.

My interest was in the Salinas Valley, [in] honoring the struggle of the farm workers. My father [Fred Farr] had done a very interesting [thing] before the farm workers ever got organized.[2] He was a social liberal and an environmentalist. We were driving down Highway 101 through the Salinas Valley to go to Los Angeles, as we did annually at Christmastime, and my mother said something to my father in the car. She said, "You know, you may be in the State Senate and [have] a big, powerful title up there, but why haven't you done anything to protect these women out in the fields who have to tinkle in the out-of-doors? It's so unsanitary, and it's so humiliating. Farm workers can harvest these crops, which we all enjoy. Can't they at least be provided some sanitary facilities?"

So my dad carried the bill in the state senate to require contractors to bring toilets to the fields so that you could have outhouses for the workers.[3] To this day, those contractors and others carrying those latrines out to the field nickname them locally as the "Farr's Chariots." I think these bills passed before the UFW [United Farm Workers] really got involved in trying to bring attention to the plight of farm workers.

[2] Senator Fred Farr was elected to the California State Senate in 1955, the first Democrat in forty-three years to represent the rural Central Coast. He held that seat until 1966.

[3] SB 899 (1965). Farr, Cobey and Stiern. Chapter 1417.

California to this day has the strongest herbicide and pesticide regulations in the United States. We also have the strongest labor standards in the fields, because we allow for organized labor. We have some strict OSHA [Occupational Safety and Health Administration] standards. You remember the fight about the short-handled hoe where we had to sue the growers. This short handle was ruining the backs of farmers by requiring them to do stoop labor all day long. No one thought you could still do the same kind of work with a long-handled hoe, which is what everybody used in their home in their own garden. The Supreme Court ruled that the short-handled hoe violated labor standards, and now I don't think you could find a short-handled hoe if you wanted to.[4]

The farm worker movement in California did a lot of things for agriculture. It made people pay attention to the workforce and to the sanitary conditions, the pay scales, the benefits, who the farm workers were. It gave them an identity. Until then, farm workers were sort of faceless and nameless. That changed, and is now obviously César Chávez's heritage. But I also think it began people thinking about, well, what is agriculture? It became broader than food. It was people, machines, water, chemicals, and climate. In the Central Coast in this fog belt—in cool climates we can grow strawberries like nobody else. We can grow [crops like] Brussels sprouts. We can grow artichokes that the San Joaquin Valley and hotter climates can't grow. We can grow lettuce. And when we grow these things, we end up growing more than anybody in America. At one time, I think almost 99 percent of all the Brussels sprouts in the United States came from Santa Cruz and Monterey County, and maybe southern San Mateo County.

We still grow forty crops in California that no other state grows. For example, all the pistachios in the country come from California. The Watsonville area, the Pajaro Valley, [was known] for apples, and Hollister was a fruit capital. Big, big fruit orchards all over San Benito County, and shipping facilities for fruit out of that area. And, of course, walnuts. Hollister has been known as the "earthquake capital" of the world, but it frankly has been the agricultural center for fruits and nuts.

Mark Lipson came to me and said, "You know, we really don't have any regulations protecting us. We've got an organic label in California, but it doesn't mean anything [without a] law. Somebody can just slap

---

[4] Sebastian Carmona et. al. v. Division of Industrial Safety, January 13, 1975.

[on] an 'organic' label and nothing can be done to stop non-organic being labeled organic." I thought he made a really good argument. I said, "If organic is really going to be organic, then we've got have some rules and regulations. Do you have any proposals?"

He said, "We got all the ideas. We know how to set it all up. We have to go out and get farmers to adopt the best management practices for organic growing. And after we've seen that they can do that and the soil hasn't had any pesticides or herbicides in it, there'll be a certification process. If they meet the test, we can write 'certified organic.'"

I said, "That sounds good to me. Is there much interest in it?"

He said, "Well, I'll bring a bunch of farmers together, and we'll come and talk to you."

I'll never forget the meeting. It was up in the top floor of the state capitol, in a hearing room. It was a very candid meeting. I looked around the room and I said, "Man, these are all hippies of the sixties, and we're up against the biggest agriculture in the country." So I said, "Look, you guys have a great idea. It's smart for California. But it's going to be hard to get this passed. I want to know how hard you're willing to work on this, because I'll put the bill in, but if you want me to just put in a bill and it gets killed, you have given the wrong message because if you're defeated, why would anybody do it again? So are you really willing to work?"

They said, "Yes, we are." I said, "Well, look, I'm a coach. This is my job. My job is to make a law. I'll show you how you do that. You go through seven gates to get to the governor, and each one of these gates has lawmakers sitting on them. They're sitting on the Ag Committee; and they're on the Appropriations, and Ways and Means Committee (as they call it in the Assembly), and then to the Assembly floor; and then you've got to go over to the Ag Committee in the Senate; and then you got to go to the Senate Finance Committee; and then you got to go to the Senate floor; and then you got to bring it back to the House to correct the amendments, adopt the amendments that were taken in the Senate; and lastly you've got to get it to the governor and he must sign it. I'll tell you who the votes are at each one of those gates, but you've got to call those people, and you got to make contact, and you have got to do it from the district they are elected from. This is a lot of work. Are you willing to do it?" They said, "Absolutely."

So I was the teacher, and they were the students, but boy, were they great. I'll tell you, the politics on it was phenomenal, because we got the bill out of the Agriculture Committee in the House, and we got it through the Ways and Means Committee, and we got it off the floor. And then all of a sudden, traditional agriculture was waking up. They had never taken any interest in this little nothing bill. All of a sudden, somebody started reading it and said, "No, this isn't nothing. This is launching a whole new industry in California."

I remember being in the Senate Ag Committee [in the California State Senate]. This is the first time the opposition had showed up. What they were opposed to was requiring that land be free of chemicals for three years prior to growing organic. They didn't want that in there, because they wanted to transition right away. Immediately. "If people are going to buy organic and we're farmers, we can't label it organic without doing this protocol. But we don't want to wait three years. We want market access right away."

The room was full because there was a lot of interest in this bill. I remember Senator Russell on the committee from Southern California, who'd gotten a message from the ag lobbyist that suggested that they take this provision out to require that the land be free of chemicals for a number of years. He made the motion, and some other senator seconded it. I said to the chairman, "Mr. Chairman, can I speak on the motion?" They said, "Certainly, you're the author of the bill." (Well, normally you don't.) [Earlier] they had asked me, "As the author, will you take this amendment?" I [had] said, "No, I'm against it." So they knew I was opposed to the amendment, but they didn't know exactly why.

So I said, "Can I speak to this motion?" He said yes. They gave me the courtesy. I said, "If this motion is successful right now, all the work that's been created on this bill and all the support around this bill will walk out of this room. This is dead on arrival. We're not carrying this bill a day more." And you could hear a big sigh in the whole room. I mean, the author of the bill had just threatened the committee that this was a killer amendment. Then the committee started talking about the motion, suggesting maybe they ought not to do it, that they ought to take my advice. They called the roll, and the motion was defeated. That was truly a great victory on that bill.

Well, when we got into the Senate Finance Committee, Senator Art Torres from Los Angeles, [who was] a big supporter of the UFW but didn't know a lot about agriculture, started taking on this bill on technicalities. I was there with Mark Lipson. This is central casting. I always kid Mark that I think he had his overalls on, which is not how you appear before a Senate Finance Committee. But [Lipson] turned to me, and he says, "I can answer this question."

So the senator laid out all his concerns, and why he thought that we ought to amend the bill to do this, because it didn't take care of this or that. Mark got up, and he said, "Senator, I'm just a farmer down in the fields, but I'm really interested in this bill, and it would really do a lot for organic agriculture. That question you asked is really a very good question. We thought you'd probably ask it. So if you'll turn to page nineteen and you look at the sentence, line seventeen, I think you'll agree that that language takes care of your concerns." And everybody went, "Whoo!" So then Torres asks his second and third [questions], and Mark answers them both like that. And finally another senator [said], "That farmer knows more about this bill than you do, Senator Torres. You better shut up." It was quite a moment. It was comic, but it was real, and Mark was just— I mean, he looked so much the part of the organic farmer. I think in politics so much you hear all these big, fancy lobbyists, but what I have noticed is people still make decisions on good, old-fashioned instincts. When you have a person like Mark, people will listen. You know he's not getting paid to get up and do this. He cares about this stuff. This is how he makes his living, and he knows what he's talking about. I've always told people that it's much better in a hearing to forget all the figures and numbers. Just tell some passionate story. Tell what this means to you. Because underneath all that crusty political titles, are fathers and mothers and people who care.

So Mark did it, and the bill passed out of committee. The governor signed it. We had to lobby him heavily, and he signed it. Everybody thought, well, this is really not going to mean anything very important in this industry. They even put the burden on the industry for the certification: "You got to raise the money to pay for the certification." [Republican governor George] Deukmejian was the governor. So we had a lot of problems. We had to put together the administration of it and the costs of that administration. But look what it's grown into.

And at that time, I found out that Senator [Patrick] Leahy from Vermont had introduced a similar bill in the United States Congress.[5] Senator Alan Cranston came to Santa Cruz. We took him up to the Agroecology Center [Farm and Garden] at UC Santa Cruz, and we

---

[5] Organic Foods Production Act of 1990.
See: http://www.sarep.ucdavis.edu/Organic/complianceguide/national6.pdf.

had a wonderful meeting there with all the folks and the students. He talked about how he was going to get involved with Senator Leahy, who was a good friend, and help him get that bill through the Senate, and he would like to be helpful to us too. Usually you don't have federal people endorsing state legislation. Cranston did, and it was news.

I tell the world that the organic movement started in California, in Santa Cruz County, and the guru of that is Mark [Lipson]. What we've done in this county is we've brought the government to the people, and we brought organics, to make them accessible. We've seen this industry take off in this region. I think its market reach is growing now at about 20 to 25 percent a year. Of all agriculture in America, the hottest niche is in organics, and I don't see it stopping, because we are now as a nation becoming really interested in nutrition. If you're going to think and act on nutrition, you're going to buy fresh fruits and vegetables, and you'll probably, given a preference, buy organic.

Bob Scowcroft

Photo: Tana Butler

Bob Scowcroft has been engaged in nearly every political development in the national organic and sustainable agriculture movement in the past twenty years—from the controversy over the contamination of apples with the pesticide Alar in the late 1980s, to the fight to pass the California Organic Foods Act of 1990, to battles over federal standards for organic certification in the 1990s, to recent lobbying efforts to secure more funding for organic farming research in the Farm Bill. Scowcroft first joined the environmental movement to work on the Alaska Native Claims Settlement Act in the late 1970s, and became a national organizer on pesticide issues for Friends of the Earth. He later served as executive director for California Certified Organic Farmers [CCOF] and led that organization through the exponential growth of the organic industry and movement. In 1992, Scowcroft left CCOF to direct a spin-off organization, the Organic Farming Research Foundation (OFRF), whose goals include sponsoring research related to organic farming practices. He served as executive director of OFRF for eighteen years and retired in December 2010.

I believe I was the first environmentalist to keynote at the Ecological Farming conference, in La Honda [California]. It was held in the winter at this Christian camp with six bunk beds per room, and the hot water ran out by the first half of the first person's shower. Everything was damp and wet and muddy. The owners of the camp, the church—they prepared the food, so it was grilled cheese sandwiches and twenty gallons of industrial tomato soup, and there were school bus homes with people living in them, too. It was the past meeting the present. This is in '81, '82, so many of the communities [communes] were still going on up there, and the painted buses were only eight years old, and the stoners, and the researchers, and the youngster farmer wannabe's—we'd [all] sit in La Honda shivering our asses off, handing out our literature or selling our books. That was really my first introduction to organic.

And their response: "Wow! An environmentalist is here! We're so far off every grid. Nobody's ever cared about our work. We're farmers, we're sustainable-ag activists, and environmentalists too. Yet the environmental community doesn't even know we exist. They're too afraid of us to have anything to do with us." Stoner, gardener, back-to-the-land—all the images that most of the major environmental groups were trying to run as far away from as fast as they could came to mind. The environmentalist thinking then was: "Oh, no. We wear suits and ties, and we're part of the system. We compromise, and we're not off any grid. We like steak and potatoes, too. Agriculture as a system really doesn't have anything to do with the environment." With the exception of a very few folks that were [concerned about] 2-4-D or 2-4-5-T or DDT, spray drift, toxic chemicals in our waterways and were reaching out to progressive farmers as allies, most environmentalists did not relate to problems in agriculture as "environmental issues."

At Friends of the Earth [FOE] a lot of things were happening to me around the early eighties. I was trying to fight spray drift of toxic chemicals and ban Agent Orange at the same time. Then Sy Weisman and Stuart Fishman from California Certified Organic Farmers introduced themselves to me by noting, "Hey, there's this representative in Sacramento [California], Vic Fazio, and his two-page organic "definition" law is about to go down the tubes as it has a sunset provision. We have a week to reauthorize it. Can you write a letter on Friends of the Earth letterhead supporting a permanent organic food act?" I said, "I'll ask." And the director and [David] Brower both said, "No, it's too extreme." I mean, we're fighting nuclear power, breeder reactors, stopping dams, saving whales, but organic was too radical an issue for FOE to support at that point in time. It was just too much.

In a typical FOE way, we spent too much time arguing and lobbying each other on issue priorities. I joined in with a request to support organic legislation. They said, "Well, okay, we'll revisit this. We have a board member whose wife is in a related field. Maybe if she says it's okay, we'll let you use [FOE] letterhead to support the law." This board member was Alan Gussow, who was a visiting art professor at UCSC and building these incredible peace poles in the Chadwick Garden, through his art classes. Joan [Gussow] was out here in Santa Cruz as well. She was a professor of nutrition at Columbia [University] at the time. She agreed to address the question of organic support. After some research she wrote a four-page letter to me, Brower, and the FOE Board, saying essentially, "There's nothing in the current literature that says organic is more nutritious. So you don't have to worry about saying it's better for you or safer because there's no data to prove it or not. But, if there is a law that defines organic, it probably will have some environmental impacts. For example if the business of organic, if the market economy of organic, has a legal framework under which to operate, then as more growers transition to organic there's sure to be beneficial environmental impacts relative to the reduction of the use of toxic pesticides and their contamination of our water."

It was brilliant. I still refer to that letter to this day. Brower and the board made the connection. The director also approved my/FOE's support and said, "Okay, you can write a letter on behalf of Friends of the Earth supporting an organic food act in California." Sy and Stuart found a newly elected first-term state representative named Sam Farr to carry the reauthorization of the CA Health and Safety Code Organic Food Act [H&S Code 26569.11], and it passed. That was my first introduction to (now) Congressman Sam Farr. Lo and behold, California had the first permanent organic law, and Friends of the Earth played a small part in that passage. Here were these organic farmers saying, "Friends of the Earth stood up for us, the only environmental group that cared," when actually it was just a frantic effort by just a few, maybe four or five farmers and allies, to see this through. But to the larger organic community at that time it was important. The end result for me was that for a number of years to follow I was invited to every Eco-Farm conference to talk as "the leading environmentalist supporting organic agriculture."

Within the same year, I was in Washington, D.C. for FOE related meetings. I had helped out a researcher named Garth Youngberg, who worked for the USDA, and planned to meet with him. He had been a professor in Illinois, and was tasked by [Secretary of Agriculture

Robert S.] Bergland to run something called the USDA Study and Task Force on Organic Agriculture.[1] They were going to organic farms, looking for local networks and calling me (among many others): "Hey, do you know any farmers anywhere we can meet?" My response was, "Well, I heard of somebody there, and there's some crazies over there, and this farmer says there's [an organic farm] in Nevada and there's one in Mississippi."

Through this grand word-of-mouth, they got to visit many farms and out came the organic task force report. It was published in the last week or two of the Carter administration. Garth was then hired to staff an organic research desk at the USDA. I was in D.C. doing something for work, when about six months later, [President Ronald] Reagan was in, and the first big, shrink-government [initiative] was to fire 500 D.C. employees in the USDA. "This is an overgrown bureaucracy"—which it is to this day, actually. They gave five hundred employees pink slips, and Garth and his secretary were among the group to be let go.

Next the order came down to destroy the organic task force report master plates, and all copies of it as well. I happened to be visiting Garth in the office the day that order came down. He was extremely upset, and I was too: "Now what are we going to do?" We decided to batch up twenty copies in a brown paper bag, and I put them in my luggage and backpack, and at that moment thought I was a law-breaking smuggler—you know, there was an administrative order to destroy this book and I'm sneaking them out of the building! I still have two of those books in my office that I kind of wave around and show at presentations.

They fired 500 people, and then brought back 498 as outside contractors. This was but one example of the bogus nature of the Reagan administration and his history of small government. They didn't shrink government. Maybe they reduced costs by taking them off the government insurance or something like that. But they needed those particular people. The only two permanently fired were Garth and this woman from the secretary pool who'd had the bad luck of being assigned to his desk months before, so she was now somehow an organic-contaminated staff person, and was let go too.

Then the medfly [Mediterranean fruit fly] controversy hit in California. I dropped everything and got involved with that. FOE thought

---

[1] *Report and Recommendations on Organic Farming* (USDA 1980).

it was a pretty bad idea to use DC-6s to spray malathion on eight million people from an altitude of five hundred feet and then follow up with a ground-spray of diazinon. They let me talk to the press and represent the environmental opposition. I'd never really done any media work. I'd never written a press release. And within three months I had the *New York Times* on line one, and the *Washington Post* on line two, and several other FOE staffers standing around me asking, "I don't know what you've done, but you're making national news. What can we do to help? Way to go!" We stopped the DC-6s, but ultimately (over the course of two years) we did not stop the spraying. However that campaign put me and Friends of the Earth on the national anti-pesticide front lines. As people then delved into our work a little bit deeper they learned that we had developed a solution to the over-use of pesticides: go organic!

Right around that time, hundreds of new local and regional citizen anti-pesticide groups were formed. Not nearly as many groups exist anymore. It's about '82 or '83— and I'm getting real tired of trying to work with these groups to ban every suspected carcinogen. They were great; it was just me. There are eight or nine thousand chemicals out there. It's going to take reincarnation over many lifetimes to get just some of them off the market. I came to the opinion that we should just get off the grid and promote organic all the way. FOE had to process my evolving political position over a year or so. It was kind of hard. There was still a lot of residual, and not incorrect [stigma around organic activism]— You know, all you need is one stoner proclaiming "herb" is the greatest organic tool going, and all your lobbying and public persona is questioned on a topic not of our own choosing.

In 1987, I saw a little ad in the local paper: "CCOF looking for an executive director," and called the phone number and said, "I'd like an application." Mark [Lipson] answered, "Yes! I've been thinking about you. You would be great," and thirty minutes later he sent the other person working there, Phil McGee, on his bicycle, over to my house to put the application in my mail box. I went through a pretty surreal hiring process and eventually got the job. There were six people on the interview committee and several of them didn't really care for each other. Issues on what the "new" organization would look like caused friction among the hiring committee. During one interview a board member comes up to me and opens his wallet to show me a membership card. "I'm a card-carrying member of the American Communist Party. Are you going to have any problem working for me?" I responded, "I don't think I've ever really had an interview question like that before." There's the card right in front of you. And he wasn't even a farmer. He sold cheesecakes out of his truck. But he was the only guy that would travel to a statewide meeting, so they [the chapter] had put him on the board.

But there were some real full-time organic farmers on the CCOF board. Warren Weber, a true leader, was president. He had assumed the position in 1985 with the goal of making CCOF a professional organization. "We're going to make this a statewide group. We're going to have real rules. We're going to enforce them, and we need an executive director that will bring it all together under the board's leadership."

In theory I'm supposed to start January 1, 1988, but I'm going into the office some evenings in the late fall just to get a feel of prioritizing what needs to be done and to get to know the other staff. Mark [Lipson], who wanted to get back to his farm, welcomed me with open arms. Others did too. Maybe too open! Over the last weeks of November and December the narrative ran something like: "Thank God you're here. There's actually a third employee, but he's not around right now and he will be leaving in March. None of us have really been paid in a couple of months, and there's hardly any money in the bank account. We're not sure what our receivables are and soon renewal farm inspections must start again. By the way, even though we were founded in 1974, and it's late 1987, we've never filed a tax return nor received our formal nonprofit status. We're really, really glad you're here, a real professional that can help us through this." It wasn't the first time I thought, why didn't I ask more questions during the interview process? I had resigned from my last position and here I was organizing my desk preparing to face the reality of no money for payroll, no tax status in place, and the 1988 renewal season ready to unfold. At one point Mark suggested we ask the Grateful Dead for a major donation. Huh?? That all came down in late '87 before I even spent a formal day in the office. Well, Mark did actually have a contact there, and *yes*, the Grateful Dead did make a $10,000 gift to CCOF through another 501(c)(3) the first week of January. Over the next month we made payroll (including back pay), paid rent, mailed out 180 certification renewal forms, and filed for our federal tax status. Since CCOF had only filed for California nonprofit tax status in January 1985, we learned that we had to file for both state *and* federal nonprofit status within three years of our original application—to the day. Of course, tax returns had to be filed during the interim application process too. January 31, 1988, was the hard, three-year deadline to get all forms legal and current.

My first formal day in the CCOF office was January 2, 1988. We had a month to complete all of this paperwork and we did it! We filed state and federal nonprofit application forms and three years of tax returns. We were fined several thousand dollars. (Thank you, Dead again.) but I then successfully appealed and CCOF eventually got its entire payment back. Soon we received our formal tax status forms from the IRS and California state government. It can be said that even though we had great staff work and board leadership, it was the Dead who saved CCOF at its most critical time. How ironic then that it fulfilled just about every image that everybody was working to get away from! And that's how we kicked it off in '88.

Warren fulfilled his obligation and left the CCOF presidency in mid-1988. An organic farmer from San Diego named Bill Brammer won the board vote to assume the presidency. Already, organic farming was gaining legitimacy. There was a state law, and more farmers were transitioning to organic by putting twenty or forty acres in CCOF's certification program. Some of the two-and five-acre growers were really upset: "Wow, the big guys are coming in. They put forty acres into organic, and there's even an organic farmer in Kern County that has one hundred organic acres. This is not good! We're the revolution. They're not, they're Big Organic." Others noted, "This is what we're all about, is everybody should be welcomed into organic certification with open arms."

The internal debate was profoundly passionate and, to some extent, size of operation remains controversial to this day. Who are we, how should we grow, particularly just how should our chapter system work? No one had ever really enforced the assessment payment scheme. The fees were based on paying a half percent on the sales of one's certified organic product. You come in, you pay a small fee to apply, and get inspected. Though you're not paying any assessment fees you are drawing on the nonprofit for services over the one to two year transition process. Then when you finally get certified you get another year of certified organic sales before you finally make an assessment payment to CCOF.

This was a real back-end problem that institutionalized instability at CCOF. We were getting more people to call for certified organic product in the marketplace, and we were getting more farmers applying new acreage to the certification program. We went from 180 to 300 growers the first year I was on staff. But even of the 180 in CCOF when I arrived, only about half of them were actually organic farmers selling in the marketplace. Many were idealists that put their forty acres of woods in certification or gardeners way into organic. They weren't farming or selling products for any appreciative income. It was a real challenge. We were always skating on financial thin ice. I'd be calling, "Could you send your assessment in a little bit early?" or "Could you guess and send in a partial payment now?" There were a few cases where I suggested that the grower wasn't providing an honest accounting of their organic sales. Confronting them with a claim of short payments was awkward but ultimately served to strengthen CCOF's certification program.

As Bill Brammer established his agenda as CCOF's president, we clicked immediately and became friends for life. He was almost seven feet tall and could be very forceful. "I've been given the charge from Warren Weber, to go forward and make CCOF a professional state-wide organization that is also recognized nationally as organic farmer directed. We have rules that everyone must agree to follow when they voluntarily join CCOF. None of you people are required by law to be here. You didn't have to join CCOF. But if you do, these are the rules you have to follow to be certified." If you wanted to stay in, you renewed and paid your assessment. That finally started to raise more money. It looked like that at last financial stability arrived for CCOF.

Eighteen months later, two things happened to change the landscape dramatically, yet again. I still stayed in touch with a lot of my old enviro-friends, and one called me one day and said, "Hey, we got a new media campaign coming down. We're going to release this report on suspected carcinogens in our food supply and do a lot of policy work around it. However we're starting to feel bad about just scaring the shit out of people all the time about poisons and chemicals so we're thinking of offering an alternative solution. I can't tell you who it is, but we are working with a celebrity spokesperson who has offered, 'Well, maybe organic food is the answer.' We don't think so.

We're concerned the public will think organic is for hippies, and we just don't want to be seen as promoting that image in the media. We trust you, and we'd like to know: what do you think?" I offered to pull together an organic briefing for this unnamed celebrity and staff, along with a list of organic farms they might want to visit. They started to come around and it eventually got to the point where they said, "Well, let me share this material with this anonymous spokesperson, and then we'll see if she wants to promote organic, because she eats it, but we're still really nervous about promoting organic." The woman was Meryl Streep. The campaign was designed to draw attention to about twenty suspected carcinogenic chemicals in a report that a group called Mothers and Others for a Livable Planet and NRDC [Natural Resources Defense Council] had released to *60 Minutes*. Meryl and another woman named Wendy Gordon had started Mothers and Others for a Livable Planet earlier that year. Both Wendy and Meryl were in favor of promoting organic under the Mothers and Others name, rather than through the Natural Resources Defense Council, because that was too much for them at the time. Though it looked like both NRDC and Mothers and Others were becoming more comfortable offering organic products as a solution, we never received confirmation that they would, in fact, support our work, either on *60 Minutes* or through follow-up media work after the broadcast.

And so [laughs]—we had one phone, an old AT&T phone in our office. It was '89, before the earthquake. When that [fallout from the *60 Minutes* report] broadcast really caught the nation's attention, the phone was really ringing off the hook, but it was still somewhat manageable. One afternoon, about two weeks later, she [Meryl Streep] appeared on the Donahue show. After listening to the problems with Alar he asked, "Now, wh-, wh-, what are we supposed to do now?" She didn't miss a beat, "What you're supposed to do is eat organic food. Go to natural food stores. Buy organic apples. Toxic chemicals are on conventional food. Demand that your store carry organically labeled products. Buy organic food." And by the time she said the word "food," our phone exploded. Right there. That afternoon. And within a week, we had a four-line system installed just to manage the incoming calls. For years after, I had saved what we called "the blue phone" with all these wires coming out of a broken plastic mold. It was my own icon of a media moment and an amazing example of the impact a celebrity can have on an important environmental issue.

CCOF expansion

CCOF grew from three hundred to eight hundred farmers and added tens of thousands of acres to be inspected in the coming six months. Our finances went completely ballistic—and off the cliff yet again.

We conducted a number of briefings after that for her [Streep] and the Natural Resources Defense Council. Apple growers were suing her personally. Alar was the chemical that for some reason *60 Minutes* focused on more than any other one. To this day, the sound bite is still called "the Alar scare," and I have to correct it every single time. It wasn't a scare. It was the truth. Alar was taken off the market. It is a suspected carcinogen, and I hear that in a few cases it is still persistent in the soil on a few farms. We were right on everything.[2] Remarkably some of the growers that sued Meryl Streep twenty years ago are now successful certified organic apple farmers.

[The second event was the] carrot caper in 1988. That was a case where this very innovative [grower] was moving these acres of carrots throughout his multi-hundred-acre parcel surrounded by cover crops, onions, potatoes and other smaller patches of organic vegetables. Imagine a rainbow quilt of organic products grown on his farm. This rotation rebuilt the fertility of his soil, confused crop-specific pests, and allowed him to present a diversity of organic products to the market-place. However, he always had carrots for almost ten months of the year. They were his major cash crop. The time came when he was out of carrots. As one of the largest organic carrot growers in California, no one could fill the market with certified product when he was out. This time around, a very large stream of organic carrots appeared from out of nowhere. They were not certified but were claimed to come from somewhere in the lower Central Valley. CCOF complained of a violation to the state government under the Health and Safety Code. We wanted an investigation and sure enough they could not provide any proof that these carrots were organic. They also said they had no enforcement power and that it was up to us [CCOF] to bring him to some sort of labeling justice. Everybody always thought it was our job to do everything related to organic. So Brammer, the board, and I talked about it, and we decided to file a formal letter complaining of consumer fraud under the organic code. Their response was

---

[2] In February of 1989, CBS aired "Intolerable Risk: Pesticides in our Children's Food," which publicized the use of the carcinogen Alar, a chemical sprayed on apples to regulate their growth and enhance their color. After widespread public reaction, the Environmental Protection Agency banned Alar in late 1989.

something along the lines of "Well, there's nowhere in the law that says we have to do anything. It's just a labeling law." "Well, if it's labeled illegally, you have consumer affairs investigate and enforce it." They still responded, "No, no way." Something about no funding to pursue consumer complaints.

At the next CCOF board meeting, on what was probably a ten-to-six vote, they approved the hiring of an attorney to go to the state and threaten to sue them if they wouldn't enforce it. His name was Barry Epstein. That got their attention: "We'll look into it but it might take a year or two to complete the investigation." Their backdoor message was that we needed a new organic food act, managed by the state agriculture agency with actual criminal penalties included in the law. All good. But we saw criminal organic fraud unfolding before our very eyes on a daily basis. We couldn't wait for a year-long formal investigation to be completed.

A woman who worked for an organic distributor (with no carrots for sale) at the time said, "Hey, I know they're doing it [falsely labeling carrots], and I found their warehouse. So I'll go in and pretend I'm writing for a newsletter and want to publish an article about their new organic carrot line. I'll take my camera, and I'll write a story on this organic business that's become quite successful and maybe I can get some photos too." She actually then went in and took shots of the long-haul trucks coming in with conventional Mexican carrots, the workers unloading them and then re-bagging them into organically labeled bags. She took these photos under some risk. We heard later that the person running the operation (it was rumored) was a wanted felon—fraudulent activity for food stamps or something like that. This was his new scam. She gave me the photos, saying, "Here they are. Here's the photographic proof. You'll have to take over from here because I can't really be part of this."

So I went to Brammer and said, "The only way I know how to push the state to act is to generate a front-page news story that will embarrass them into immediate action." We brought a plan to the CCOF executive committee: "Here's the risk of seeking publicity, publicly pointing out that this "organically labeled" product is associated with cheating and fraud. If we succeed it will also become apparent that there's no real organic law enforcement. Some may declare that organic is just a fad and not worth the money. The flip side is that we may really bust him. We'll then succeed in multiple ways. We embarrass the state and we come out saying, 'We're tough and we're not taking this crap anymore. If the state won't do it, we will.'" It was a very hardy and vigorous debate, and all understood that in some ways CCOF's very existence was at stake. The end result was, "Approval for action. Yes, Bob. You go and release the photos. You and Bill serve as our spokespersons. You seem to know what you're doing, you know the press, and standing up for our certified organic membership is worth the risk. What worth is it to be certified organic if we know there's fraud?"

Several years earlier I had met a reporter for the *San Jose Mercury News*, which was a very powerful paper at the time before these days of media consolidation. I drove over there and pitched, "I got a story that'll put you on the front page. I'm sure of it." He was a great reporter, young but aggressive. He said, "Well, this looks real, and it's a good story." Over the next week or so we went back and forth on the contents of the story. For example, why did the Health Department cover organic labeling, he wondered, and not the Agriculture Department? And his editors were really concerned, in particular, that they didn't have permission from the people in the photos to print them in the paper. Asking permission might have defeated the purpose, I responded. We had any number of at-home, late-night phone calls as the story prepared to go press. At one point he called, "My editor has another problem. I got three hours left. He's pulled it due to one more unanswered question. Can you help now?" I did. Mitchel Benson was the reporter's name, and the *Mercury* put his story on the front page along with the photo of the worker holding the organic bag and packing it with boxes of Mexican conventional carrots.[3] Within seventy-two hours, the Associated Press had picked up the story, and about twenty other papers called CCOF for additional comment. We said [to the media], "You call the state and ask them why they didn't enforce the labeling law. We filed the complaint months ago. Here's our attorney's name. This has got to stop right now."

Camera crews went down to his warehouse, and maybe for about a day, he [the packer/shipper] said, "These people are just trying to put me out of business for their own special interests." Then he disappeared, just fled, and the state was justifiably embarrassed. They

---

[3] Mitchel Benson, "Carrot Crisis Organic Veggie Scam Alleged," May 11, 1988, Page 1A, *San Jose Mercury News*.

accelerated their investigation and agreed that these carrots could not be labeled organic. This event put CCOF on the national media map as spokespersons for organic family farmers. The state also noted, "We are not going to conduct any more enforcement actions. We have no budget, no money, and no staff that understand organic farming." CCOF faced the new challenge head-on. At its next board meeting, it voted to hire Barry to work with Mark Lipson to write a new, enforceable organic foods act for California. Sam Farr agreed to sponsor the legislation. Darryl Young was his legislative assistant who worked with Mark Lipson and Barry Epstein to move it through the state legislature. AB 2012 was passed in 1990, thanks to CCOF's first ever lobbying effort, major support from our environmental allies, and a number of nights sleeping on the office couch in Sam's office in Sacramento.

The original California organic law contained language allowing a one-year transition to organics, while Oregon and Washington had a three-year [transition] period. When word circulated that we [CCOF] were going to write a new law we started meeting with Oregon Tilth, Yvonne Frost and Lynn Coody in particular, and Miles McEvoy, Margaret Clark and Gene Kahn in Washington Tilth.[4] Soon we made a handshake agreement that if California went to three-year transition instead of one, they would agree to merge Oregon Tilth's and CCOF's [approved organic] materials list. With a general set of principles in hand we needed to get down and write the fine print. We knew that eventually all three state laws must be synchronized to allow the cross-state-line organic trade to expand. Gene Kahn offered, "Well, Cascadian Farms is expanding and I want to give back. I'll contribute funds for meetings, airfare, and printing costs to cover most of your expenses." Each of our three nonprofit organizations donated staff time and some funds as well.

We were ready to go. We called ourselves the Western Alliance of Certification Organizations, or WACO [pronounced "wacko"]. We had no budget, no staff, no minutes, no formal meetings, just a great sense of humor and the moral high ground. We held special meetings three to five times a year, usually taking advantage of regional conferences, for about two years, and we sat down and wrote legislation. We

Handwritten margin notes: "No $ for State enforcement", "Certifier", "Regional Cooperation"

[4] Oregon Tilth and Washington Tilth are organic certification organizations. Their histories both reach back to the 1970s and the formation of Tilth, a regional network of organic farmers and gardeners in the Pacific Northwest. See http://www.washingtontilth.org/history.htm

believed that this is citizen advocacy at its best and since anybody can do it, it might as well be us. We had a good dose of attitude. I was one of a few, maybe three to five others, that had a combination of legislative, lobbying and grassroots organizing experience at that time. I was a strong advocate, thanks to backing from Bill Brammer and the CCOF Board, to go for it: "You'd be amazed. People in state legislature get so little mail, so few new ideas and most likely have never met an organic farmer. Let's just walk in there with an attitude and say, 'We're here to write an organic food act and we need your support.'"

It doesn't seem possible now, but as the carrot caper was making national news and the WACO initiative was taking shape in three state capitals, an aide for [Vermont] Senator [Patrick] Leahy called the CCOF office informing us that the NOFAs [Northeast Organic Farming Association] were asking for state organic dairy regulations in Vermont. She had heard there was organic legislative activity in California and wanted more information. Since Mark Lipson was going to Washington to attend a conference, I asked him to go into Kathleen Merrigan's office and tell them we need harmonized standards between all of the states. We need a national law. What the hell?"[5] [laughs] When Kathleen tells her story, it was this guy, Mark Lipson, who just walked in and said, "Well, I'm here from CCOF, and we need a national law. The three Western states will soon have legal standards. But there's mislabeled, even fraudulent organic product coming into the West from states that do not have an organic labeling law. They're beating us up. There is no national means to enforce certified organic products." That was the day conversation began on writing the [Federal] Organic Foods Production Act, which was approved by 1990. What a time that was.

---

[5] At that time, Kathleen Merrigan was a senior staff member of the U.S. Senate Committee on Agriculture, Nutrition and Forestry, working for Senator Patrick Leahy (VT). In February 2009, President Barack Obama chose Kathleen Merrigan as deputy secretary of agriculture, the number two position in the United States Department of Agriculture.

Mark Lipson

Photo: Tana Butler

Mark Lipson was California Certified Organic Farmers' first paid staff member, where he worked from 1985 to 1992, steering the organization through the establishment of a statewide office as well as several key historical events that awakened the American public's interest in organic food. He served as senior analyst and policy program director for the Organic Farming Research Foundation (OFRF) from 1995-2010. In 2010, Lipson took the position of Organic and Sustainable Agriculture Policy Advisor for the United States Department of Agriculture. The Organic Center calls Lipson "the primary midwife" of the California Organic Foods Act (COFA) of 1990, authored by then-State Assemblymember Sam Farr. Lipson is also a partner in Molino Creek Farm, a cooperative farming community located in the hills above the ocean near Davenport, California. Molino Creek pioneered the growing of flavorful, dry-farmed tomatoes (grown without irrigation).

In 1977, I applied to UC Santa Cruz and got into the environmental studies program. It was kind of a golden age for the environmental studies program [the late 1970s and early 1980s] because it was very real-world oriented, and the students were out all over the place, all over the world, doing practical stuff. For example, in 1978 we had the first prospect of there being offshore oil drilling off the Central Coast here, so with a couple other students, I got into an internship with AMBAG, the Association of Monterey Bay Area Governments, preparing AMBAG's comments to the Department of [the] Interior on the first call of the question for offshore oil leases. Right from the very beginning, there was this assumption of doing stuff in the real world. There were no purely academic exercises. I went into the Planning and Public Policy track [in environmental studies]. I had interests in political organizing, had read some Saul Alinsky.

The other thing I got into was co-op organizing. It was also in my second quarter. I was taking a class on [urban] planning, and we started this discussion about housing and housing co-operatives, and why wasn't there a student housing co-op in Santa Cruz? It was and is a big feature of Berkeley and other places, but there was not anything like that in Santa Cruz. The next quarter I taught a student-directed seminar with a couple of other ES students to develop a plan for a student housing co-op. Co-op organizing was a major theme of my college career. I got involved with Neighborhood Food Co-op and with some statewide co-op networks.

I also got into Molino Creek Farm basically through the co-op window, because it was being organized as a co-operative. We were starting up our farm and that led me to California Certified Organic Farmers. We were wondering about organic certification; we knew that there was such a thing, but didn't really know anything about it. So I found CCOF and Barney Bricmont somehow and started going to a couple of meetings of the Central Coast chapter. This was '83, '84. They had just begun to tax themselves 1/2 or 1 percent of sales to create the fund for—I think the original idea was for marketing, for promotion—but it also created the possibility of starting to have a paid staff and establishing a professional office. I put myself forward for a position called executive secretary, which was a part-time deal.

CCOF was growing. There was this watermelon incident in 1985, July Fourth. A bunch of people all around the country got sick from eating watermelon that had been contaminated with aldicarb, with TemikR insecticide. That was the first sort of hyper-drive accelerant to the growth of CCOF, a pesticide scare in the summer of '85. At that point, CCOF probably went from eighty or ninety members around the state, maybe grew by 50 percent pretty quickly. The curve was rising. We were starting to do inspector trainings and writing the first certification manual to kind of codify everything. Things were cooking along and bringing in more income to the organization, new chapters forming in some places in the state. So the board made a decision to hire a full-time executive director.

This would have been early '88. I didn't want to be the full time ED. I was happy with my part-time position and continuing to farm at Molino Creek. I went to New Zealand to go to the—I think it was the second International Permaculture Conference—and to do a permaculture course with Bill Mollison out on this little island off Auckland.[1] I was in Hawaii on my way back from that trip when the Alar thing hit.[2] Things just went immediately, totally ballistic. There was this big *Sixty Minutes* exposé about Alar. These things (pesticide contamination events) had been building for a few years. It just went over the top completely.

I was coming back from a very different experience in New Zealand. I mean, Mollison completely blew my mind. I was going to quit my job and just permaculture Molino Creek. That's what I wanted to do. And that went out the window really quick when I came back to this mayhem. I mean, the growth curve of CCOF and the media attention—you just didn't know what was going to happen next. We had a front-page article by the environmental writer for the *New York Times*.[3] It was this profile of CCOF, saying maybe it was the most influential alternative

[1] Bill Mollison is an Australian ecologist, teacher, and writer who is considered the father of permaculture. "Permaculture — from permanent and agriculture — is an integrated design philosophy that encompasses gardening, architecture, horticulture, ecology, even money management and community design. The basic approach is to create sustainable systems that provide for their own needs and recycle their waste." See http://www.scottlondon.com/interviews/mollison.html Also see: http://www.permacultureplanet.com/

[2] "A watershed event in the consumer food safety crisis was the release of "Intolerable Risk: Pesticides in our Children's Food" by the Natural Resources Defense Council (NRDC) (Sewell et al., 1989). The report attacked procedures used by the Environmental Protection Agency (EPA) to estimate health risks from pesticide residues and the time taken to discontinue a pesticide's use once it is found harmful. At issue was a breakdown product of the growth regulator daminozide (trade name: Alar), which is sprayed on apples to prevent preharvest fruit drop and to delay fruit maturity and internal decay. A *60 Minutes* broadcast and other media coverage about the report created nationwide panic. Although government and other scientific experts refuted NRDC's charges, public outrage eventually resulted in Alar's voluntary removal from the domestic market by Uniroyal Chemical Company." M. Elaine Auld, "Food Risk Communication: LesHealth Education Research Volume 5, No. 4, 1990. pp. 535-543.

[3] See Marian Burros, "The Fresh Appeal of Foods Grown Organically," January 28, 1987. *New York Times*, Pg. C1.

agriculture group in the country. We were the most professionalized of all the organic groups around the country. There were these grassroots, bootstrap grower-based certification groups, all spin-offs from Rodale, basically. That's where the genesis of all of it was, was Rodale.[4] That's how CCOF had gotten started ten years before I was involved. But CCOF was the most professionalized, and had a real office and a phone, and [our new executive director, Bob Scowcroft] knew how to run with this stuff, and what it meant to be an activist, professional organization in the spotlight, so things really started growing fast.

In the wake of the Alar episode, there started to be a lot more interest in organic. Of course, California had had a law on the books,[5] but with all the renewed interest, there was starting to be some focus on that in the Ag Committee of the state legislature. Sam Farr was in the State Assembly. There was an exotic-pest crisis with the apple maggot, and some of the organic growers up in Humboldt and Mendocino County were getting kind of militant about their rights not to be sprayed and their economic interest in being organic. So the guy who was the chair of the Assembly agriculture committee, I can't remember his name— an established, typical state politician, big friends with big ag—started making noises about overhauling the state organic law, and it wasn't necessarily going to be a good thing.

I started making trips to Sacramento as a CCOF employee, and started working with Sam and his staff on what turned out to be a major overhaul, a complete, total overhaul of the state law, and moved part of it out of the health and safety code into the ag code.

At the same time, I helped instigate the federal law. I had gone to Washington for a meeting of the National Coalition Against the Misuse of Pesticides, NCAMP (which is now called Beyond Pesticides), and was cultivating this crossover between the anti-pesticide groups and the organic farming groups. NCAMP was a national group. It was centered on the people who had immune disorders from pesticide exposure, the

---

[4] J.I. Rodale founded Rodale Inc. in 1930. In 1942, he started *Organic Farming and Gardening* magazine. Rodale is recognized as the father of the natural food and natural living movements in the United States. See: http://www.rodale. com/1,6597,1-101-211,00.html. See also: "Oral History Interview with Robert Rodale," by the Alternative Farming Systems Information Agricultural Library, NAL Call No. Videocassette #670.

[5] Lipson is referring to the California Organic Food Act of 1979.

environmental sensitives, and the hardcore anti-pesticide activists, which didn't necessarily include farmers.

While I was in Washington on that trip—again, post-Alar, a lot of media attention on this nascent organic thing—there had been a proposal floated in the U.S. Senate Ag Committee to create a federal standard. It came from Georgia, Senator Wyche Fowler. This was not being dictated by pro-organic interests. So I made a cold call to the staff of the Senate Ag Committee. There was a brand-new person in that job named Kathleen Merrigan[6], and I proceeded to tell her about organic agriculture: "You know, there're rules in a few different states, and they're all kind of different, and [we] probably need to start thinking about a national standard." So that was sort of the inception of what turned into the Organic Foods Production Act of 1990.

I made the first snowball that turned into that avalanche, but I was mostly focused on the California state legislative process and overhauling the state law. Those two things (federal and state laws) were moving simultaneously. A big national network of organic growers got put together to work on the federal law. United Fresh Fruit and Vegetable, which was the big national mainstream produce promotion organization, helped convene some meetings that brought together a lot of the organic folks. We had a network of organic grower certification groups called the Organic Farmers Association's Council, which grew out of WACO, [pronounced "wacko"] (laughs) which was the Western Association of Certification Organizations. That was CCOF and Oregon Tilth, and Washington Tilth. Because this issue of different standards really was a problem, we were trying to iron all that out, or at least connect with each other to try and figure that out. In 1990 almost nobody had email. How did we ever get by without email and cell phones, I don't know.

The thing that really sparked the certification issue in California was fraud, the "carrot caper." This is what really kicked it into gear—these people were selling organic carrots that basically were fraudulent, but there was no way to deal with it. The retailers said, "Certified? Uh, no, they just told me it was organic, so I labeled it organic." The distributors were like, "Yes, I need some organic. Go ahead, call it organic!" There wasn't even a basic acceptance of third-party certification as a baseline requirement for the marketplace.

---

[6] In February 2009, President Barack Obama chose Kathleen Merrigan as deputy secretary of agriculture, the number two job in the United States Department of Agriculture.

The [higher price for organic food provided] a very attractive incentive for fraud. We [CCOF] made a deal out of one of these cases in the media. The *San Jose Mercury News* ran with the story.[7] The basic objective became to create an enforcement program, to create an assessment collected by the Department of Agriculture and the county commissioners, and have a registration program. [At that time] the board of CCOF didn't feel like it was right to impose a requirement for certification[8]. They wanted [certification] to be [an] additional level of value in the marketplace. But they said, "There needs to be some kind of legal baseline, so we'll create this registration program where everybody has to basically declare that they're following the law and make a legal affidavit that they're following the standards in the law, but then, third-party certification will be over and above that." That was the point of view of the CCOF leadership.

Meanwhile, the federal law was being written to require third-party certification. Of course, that took over a decade to actually come into force. But the California legislation itself was a very epic process. This was at the time of Alar and the cyanide in the grapes.[9] Pollutants in the environment and contamination and carcinogens were [a] big issue. [There was this] huge, comprehensive, totally game-changing proposition called Big Green [Proposition 128 of 1990]. It would have really directly affected agriculture. So a lot of mainstream agriculture was looking at all this saying, hey, in a few more years, we're not going to have any chemicals, and everybody's going to have to be organic. So everybody in Sacramento became intensely interested in this bill that Sam Farr was moving along to upgrade the organic statute and create a regulatory program. We started having these meetings with thirty people, and all the major ag organizations, and their lobbyists, and their mother's lobbyist, (laughs) and you name it! Five or six different state agencies would be sending people, and it was quite a tizzy. That confluence of what had happened with Alar and other stuff in the food supply, and the overall environmental battles going on really intensified it all of a sudden.

---

[7] Mitchel Benson, "Carrot Crisis Organic Veggie Scam Alleged," May 11, 1988, Page 1A, *San Jose Mercury News*.

[8] Lipson means that in order to be able to label their produce organic, growers would have to have certification.

[9] See "Don't Eat Grapes, FDA Warns; Cyanide Traces Found In Fruit From Chile After Phone Threat," *Washington Post*, March 14, 1989.

This was all going on around the time of the [1989 Loma Prieta] earthquake. [CCOF's] office had been destroyed. We had hired this environmental attorney in San Francisco, Barry Epstein, who did a lot of work in Sacramento, so I was working out of his office in midtown San Francisco, and every time the bus went by, the whole building would vibrate. I was in post-traumatic shock. It was very nerve-wracking. And I'd be driving up to Sacramento. I spent nights on the couch in Sam Farr's office in Sacramento. It was all very, very intense. But we got both of those bills passed, state and federal, and then went into a whole new phase.

The Organic Farming Research Foundation was created by CCOF in 1990. It was very clear the universities weren't helping on organic, and the growers' attitude was, well, we're just going to have to do this ourselves. OFRF was created as a vehicle for funding research. The idea was not based in any positive expectations about the universities. It was, we're going to fund farmers to do the research and share it with each other.

In trying to grow OFRF, there was interest from some of the funders in public policy related to research. Policy work wasn't really one of the purposes that OFRF was formed for, although it was within the scope of its mission. But Bob [Scowcroft, founding executive director of OFRF] and the board were seeing interest in this from some of the foundations that they were querying about funding. So we decided to put together a proposal that would fund me to become a part-time staffer for OFRF to examine federal research portfolios. We got funded by the C. S. Mott Fund to do this analysis of [the] United States Department of Agriculture [USDA] organic research portfolio. The study [entitled "Searching for the 'O Word'"] proved to be a much more effective platform for policy advocacy than we really had any idea that it would be. It had some hard data [on how little funding was being provided for research on organic farming]. It put some real analysis into something that was otherwise just anecdotal, and it coincided with the first proposed rule for the National Organic Program, which was very much caught up with the early Internet phenomenon. When that rule was published [in December 1997], it was one of the first rules that you could comment on electronically. It received more comments, over a quarter million, which was an order of magnitude more than anything USDA had ever done, and the second-most that anything in the federal government ever had received in terms of public comment. The only other thing that got more was the FDA's attempt to regulate tobacco as a drug. This is just absolutely astonishing, this seismic event that wasn't even necessarily appreciated by [the organic farming] community, but within the government was recognized as something really, truly astonishing.

We had a really effective national coalition. The movement had grown— there were organizations all over the country. And the market was really growing. There was a critical threshold of legitimacy, or realness, concreteness about organic that—it really can be done. It's not just a hypothetical or a pipe dream; it's really happening. By then, people had been doing it for ten years, and were growing and economically viable. Some scientific stuff was starting to accumulate and always ever more stuff about the dangers of pesticides and synthetic fertilizers. All those things contributed to this intense interest in it. There's no question in my mind that political response to the first proposed organic rule was the first exercise of the new muscle tissue of what we look at today as this massive movement for local food.

The coincidence of "Searching for the 'O-Word'" with that political [outcry] immediately helped the "O-Word" document get a lot of attention. People were slapped upside the head in Washington, saying, "Organic? What's organic? And what are we gonna do about it?" (laughs) And so behind that shock wave, the "O-word" [study] was there. We gave it to people in Congress. And so [California Congressman Sam Farr, who was not brand-new to Congress but pretty recent, introduced a provision into the research title of the 1997 farm bill, except they did the research title separately a year later. It was actually the 1998 Agricultural Research, Extension and Education Reform Act [AREERA]. They booted the research title an extra year at that time, but Sam established this unfunded provision for an organic research program. It didn't get any funding until 2001.

The essential policy problem right now is that the growth of domestic production of organic is not anywhere near what the growth in demand is. The market keeps growing by 20 percent a year. Sales of organic in the U.S. just keep charging ahead, but that is not translating into conversion of land to organic and conversion of farmers to organic. It's a very complex picture. It's different in different regions of the country, and it's different for different types of agriculture, the obstacles. We think the number-one limiting factor is lack of information—lack of both specific research results in how to build organic systems, and a lack of the extension and training capacity to help farmers actually access whatever information there is.

It's just within the last few years that [organic] research has actually started to get in the pipeline. You have to have the research, and you have to have a specific capacity to utilize and transmit that research, and neither of those are very robust. They're still in their infancy. While the market is charging ahead, there are still these just little tiny baby steps happening in the research and extension system. We need to seriously ramp this up, or else this market and all its benefits—they're going to accrue to producers in Argentina and Mexico and China. And there are new proposals specifically to assist organic conversion, financial incentives. Financial assistance.

This year [2007] has been really, really remarkable. Organic has arrived at a place that still is quite out of proportion to its actual presence in the landscape, but it reflects the potential that it has and the background it stands out against. The groundwork that we've been doing for fifteen, twenty years is paying off. The careful nurturing of credibility and respect is paying off, and we're doing the best we can, approaching geezerhood, to be nimble enough to take advantage of it. We're playing a role in the national scene, with a high level of responsibility, trying to integrate organic with all these other things that are happening—local and food security and conservation. It has its mainstreaming aspect. In some ways organic isn't enough, or good enough, or radical enough for some people. They want "beyond organic." My response to the phrase "beyond organic" is that we haven't even gotten to organic yet. I mean, it is still not anywhere near being fully realized. It's still in a rudimentary state. The whole regulatory thing is only partially implemented. We barely have a trickle of research and development happening. So it's just a fallacy to say, "Oh, well, organic's not good enough; let's get beyond it," because it's still very, very much a work-in-progress.

You know, watching C-SPAN you still hear people blather about America feeding the world and that is one of the stupidest lies, really. America is not feeding itself. Yes, we ship a lot of grain around the world, but the U.S. has been a net food importer for several years now, and that's going to get worse really quick. And in addition to that, the whole infrastructure of the U.S. food supply is getting more and more fragile. You have food safety issues. You have labor issues. You have tremendous energy cost issues, which ethanol is not really going to get us out of. The cost of food is rapidly escalating. There's a big article in the latest *Economist* magazine about the end of cheap food, and it's been picked up by a lot of columnists and writers in the last couple of weeks. The last five years have seen a change in the global cost of food that is radically different from the last hundred-some years. *The Economist* magazine has a food cost index that goes back to before 1900. It's just like this really flat line all the way along. In the seventies it starts bumping up and down, and then in the twenty-first century it's just goes straight up. That's going to be a permanent feature of the landscape. Then there is the cost of oil and natural gas for making synthetic fertilizer. There are so many points of vulnerability for the system that we have. So whether or not we can cultivate a robust, more regionalized food system before the rest of it falls apart, I don't know.

There's a lot of dispute and sort of retrospective debate about organic: "Is what's happening now part of our vision that we had at the beginning? Haven't we betrayed our roots and let it become corporatized and owned by Wall Street? That isn't what we were doing when we started all this." I think a lot of that is very muddled, and a lot of the people who talk about that really weren't there at the beginning either. But one thing that very few people do talk about that was, I think, at the root of what a lot of people had in mind when they went back to the land, or thought they were going back to the land and then tried to start farming, was this idea of a fallback, that this civilization is not going to be able to sustain itself and organic agriculture is preserving some kind of capacity to withstand the shocks of the large-scale industrialized system being disrupted. I think that is starting to come back a little bit.

For the last hundred years we've had this one-time free bonus of all this petroleum to produce synthetic nitrogen fertilizer. When that goes away or declines, how are we going to produce food? We're going to have to have something that is rooted in what organic agriculture is now. Organic agriculture is still very rudimentary. We are still at a very rudimentary stage of being able to have really productive, resilient, nontoxic systems. We're doing it with a lot of Band-Aids and crutches that are borrowed from conventional agriculture. The research and development input into organic just has never been significant. So comparisons about—well, you couldn't feed the world organically, or we couldn't survive economically organically—well, the system that we have has had billions of dollars and thousands of scientists working very assiduously on it for many decades. And that input is what has produced the success of the system that we have such as it is. So what could be the potential of organic systems if they really had that input of science and development? We haven't even begun to imagine that.

Tim Galarneau

Photo: Jennifer McNulty

Farm to College et CASFS

In its March/April 2009 issue, Mother Jones *magazine called Tim Galarneau "the Alice Waters of a burgeoning movement of campus foodies." Galarneau is a cofounder of the Real Food Challenge, a national campaign promoting sustainable food-sourcing in college dining halls. In his day job with the Center for Agroecology and Sustainable Food Systems (CASFS), he coordinates the Center's Farm to College project. Since his undergraduate days at UC Santa Cruz, Galarneau has helped spearhead numerous initiatives to change the way the nation's schools, hospitals, and other institutions navigate the high-volume acquisition and preparation of food. Galarneau also was a leader in transforming dining services at UC Santa Cruz. The campus now contracts for organic produce to serve in its dining halls with a consortium of local farmers. Carefully developed purchasing guidelines not only prioritize the direct acquisition of local, organic food, but also emphasize equitable labor relationships, environmentally friendly farming practices, humane animal husbandry, and a university food service that is as much about education as about feeding a hungry campus population.*

I grew up in a home that had fast food; we had microwaveable dinners. We had traditional American meatloaf and casseroles, and the meat was cooked until it was of shoe-leather quality. I was excited about it only if there was ketchup involved. Iceberg lettuce and the whole nine yards. So coming to California, I tasted my first bit of tri-tip from an oak grill, at medium rare. I didn't even know I was tasting meat, it was so incredible. It made me begin to think about my food choices.

Stepping out and buying food for myself, thinking independently, I began to question a lot. And then I began to read: First, about the American dumbing-down effect and the disenfranchisement of indigenous cultures and knowledge from Jerry Mander in his *Four Arguments for the Elimination of Television* and his *In the Absence of the Sacred*. Then about water wars, and *Stolen Harvest*, and issues of genetically modified organisms from Vandana Shiva. Then beginning to look at food choices in particular.

When I went to Santa Barbara City College, my first goal was to understand why people make the choices they do. What are the childrearing practices, the cultural practices that impress children to shorten their sense of themselves and their ability to make choices and think independently? So I focused on psychology. I joined the honor society and I did a lot of altruistic outreach events—food-bank fundraising and food drives for the local food banks and the boys and girls clubs.

It was really the [transfer] to UC Santa Cruz, in 2002-03, that focused [me] on food issues. I heard there was an annual fall festival where all of the organizations let students know of their opportunities. A friend of mine was behind the table of the Student Environmental Center, talking about transforming the world through sustainability and collaboration. They were working on solar energy and green building for the UC system; they had a campaign called Students for Organic Solutions that was just launched, as well as a waste prevention campaign.

I went to a fifteen-minute breakout with Students for Organic Solutions. I was hooked with that. I had been exposed to potlucks and community dinner gatherings while I was in Santa Ynez that I hadn't heretofore experienced, and realized, wow, this is very powerful, to bring people together, so I also instilled that in my home life at [UC Santa Cruz's] College Eight. We'd hold a weekly potluck for the local residents and build community.

During that whole year, we were trying to build connections with UC Santa Cruz's contracted food service provider at the time: Sodexo, Marriott Food Services. And Sodexo, during our meetings with them, wasn't that optimistic about bringing in any—let alone SOS's goals of 100 percent organic food in the next couple of years.

Of course, the students were very ambitious to want to see our values met in real time in the real world. We hit some hard challenges and hurdles. Barriers before us from Sodexo were, "We don't work with local vendors. We work with our preferred vendor list, including with produce. We have a three-to-five-million-dollar insurance liability waiver for any farm or producers that want to work with our contracted food service operation on this campus. Beyond that, we haven't at this time sourced organic or local. We haven't noticed it as an interest or demand."

Hence a survey as an initial step. At that time, the Center for Agroecology and Sustainable Food Systems was offering grants for students to propose research projects. Now, I wasn't an environmental studies student. I was a psychology student, and then soon to be also a community studies double major. I wanted to do a campus-wide survey, a student organic awareness survey. I wanted to develop a method and process that could be done online and to get a random sampling of students. I proposed this project and they funded it, which was really exciting. The process carried on the whole year—of setting the study up, confirming it, getting it launched, with incentives of prizes and so forth. We had great feedback and input to bring to Dining Services.

But along the way we realized that there's more to the situation than the contracted service vendor. There are campus representatives that have oversight. And so in a power-mapping of the process, we realized that we needed to talk to Alma Sifuentes, who at that time was the Director of Residential and Dining Services.

So we met with Alma. We let her know we were doing outreach and educational events to the students, that we were doing this survey, and we wanted to know, "What type of pressures could you exert on Sodexo so that they would be interested in offering product that builds the value and integrity of our sustainable and regional food and farming economy and lifestyles and culture?" And Alma offered, "Get us the results. We'll continue this dialogue."

At the same time, during this year there were concerns about labor. The labor unions were challenging the double standard of contracted food service workers who were making two dollars an hour on average less than service workers who were UC employees. [The contract workers] had no access to health care benefits for their family, nor the benefits of being a university employee, so it was a very degraded position. There were management/labor issues and tensions throughout the year.

There was a renewal period [for Sodexo's contract] coming up that spring, so there was almost like a perfect positive storm. A group entitled Students for Labor Solidarity contacted Students for Organic Solutions, having heard we'd been in touch with Dining, and wanted to know our positions. They let us know that they were going to launch a

"Dump Sodexo" campaign that would get petitions to raise awareness, to do negative press positioning on Sodexo and its relationship to the university—escalating to Valentine's Day, where they would march to the chancellor's office and let her know the concerns of the campus community.

So what we had was like this mini-UFW uprising—with students, labor groups, concerned community representatives on the campus wanting to take a position on this issue. So we marched. During this same time—it was such an exciting time!—the United Students for Fair Trade, the national youth initiative, was forming and hosting a convergence in Santa Cruz overlapping with Valentine's Day. Students for Organic Solutions had partnered to help host a Valentine's dinner event on the Farm. "Where's the love in our food?" we called it. [laughs]

So we hosted this dinner event on behalf of the Student Environmental Center and our campaign, and we were able to bring people together after this march and rally that day, and had a phenomenal dinner, and began to look at the intersections of trade and justice and youth involvement in activism, and at SOS's role, as more than the environmental issues of our food but also social equity. It was a very powerful learning experience.

And two weeks later, the university issued a termination of a thirty-two-year contract with Sodexo. The scale of the campus would allow them to develop an in-house system, so that they could run their own food service operation and bring all the employees on board as university employees.

I chaired our Student Environmental Center the following year, after coordinating the Students for Organic Solutions campaign. I learned about hosting and coordinating our annual Campus Earth Summit with a team, where we had no funding at the time. We would bring everyone collectively together from the campus—staff and faculty administrators, the NGOs and community representatives—to target focused discussion groups on topics of sustainability that mattered and were in the minds of those in attendance. We were practicing a very interactive meeting design, versus a traditional Powerpoint conference session format. The goal was to bring ownership into the process from the people most invested and empowered to do change on the campus.

Within three years, we saw a transition from 0 to 27 percent local organic food sourced from socially responsible farming operations served in campus dining facilities. With over 14,000 meals served a day, this impact for our [local] growers is phenomenal.

I'm also inspired by how Dining Services has taken these ideas of sustainability and green, and they've run the whole gamut. They signed a code of conduct to offer sweatshop-free uniforms for food-service employees for dignity in the workforce. Their contracts did increase their wages by over two dollars an hour, and offer health care benefits for their families, and access to the library, and the health/wellness centers—all the resources that a university employee should have. They're doing trainings in how to reduce waste, installed pulpers that reduced waste by two thirds and the rest is a compostable material. [They're moving] towards recycling their oils for Salinas Tallow and their bio-fuels unit. They are making bridges and connections and thinking in a systems fashion.

We succeeded in including a worker-supportive criterion in the campus food services contract with its local organic suppliers. I think that the idea of social justice for bureaucracy and administration has negative ramifications due to their experience with direct intense union dealings. When they think social justice, they think unions, and when they think unions, I don't think they necessarily think equity. They think, another bureaucracy, a mini-bureaucracy they have to then answer to and work with. So when we wanted to put social justice criteria in contracts, the first thing we heard from [UC Santa Cruz] Purchasing was, "Well, it's not a food attribute, so it could not legally stand in any way. Secondly, there is no third-party verification of the larger integrity. Unions may offer prevailing wages, but they might not cover the whole gamut of socially just. And it's open to discernment. There's no third party certifier for that." We had no domestic fair trade and social justice certification.

So a compromise was reached. In contract negotiations, you have a cost point plus system. For favorable attributes, you get points. The better overall score, that's how you're weighted for whether you're going to get a contract with the university. So we were able to put in preference points for worker-supportive—which includes educational capacity building, subsidized transportation and housing, health care, union or prevailing wage.

The really interesting thing was we were able to get *local* in there, and *seasonal*, and *direct*, as [criteria], so we could put this into binding contract language. And the cost issue, which was, I think, very important in this process. If we went to a traditional lowest-common-bidder process, those points would rate high for big, industrial agribusiness food providers: your U.S. Food Service, your Syscos, and so forth. A consortium of growers wouldn't ever be able to manage to draw up and lowball a contract, and maintain it, because they don't have the diversity of product they're [Sysco et al.] selling.

What we learned was there is always an answer if you look for it long enough. [In this case] it was a process called sole sourcing. It allows the university to develop an exclusive contract based on reasons that do not apply to traditional vendors. Your traditional vendor can't provide a [particular kind of] product [that the institution requires], so you have an opportunity to sole-source. You only need [to fulfill] one [of the criteria], but I believe there are six. We had two: First, there is a unique research and educational connection; we were doing research and education with local growers. The second was that it's a unique product that would not be available through the traditional supplier. The traditional supplier could not guarantee that [their product] would come from within a 250-mile radius, or that it would be picked and harvested that same day. They could not guarantee the freshness because their sourcing was part of a global agro-industrial food system.

A sole-source contract allowed us to sidestep a traditional bidding process and convene and develop a direct contract with—one of the most outstanding achievements on the campus—a consortium of organic growers, the Monterey Bay Organic Farmers Consortium. It includes Happy Boy Farms, Coke Farms, Phil Foster Ranches, Swanton Berry Farm, New Natives Farm, as well as the Agriculture & Land-Based Training Association (ALBA). So we're talking about working with long-time, committed, environmentally and socially conscious producers, and innovative programs, such as ALBA's, that take aspiring farm workers and low-income farmers and disempowered individuals and empower them with the tools for becoming organic farmers and developing their markets. An amazing combination of people came to the table. We emailed to over a thousand farmers through CCOF's and CAFF's networks, letting them know about this meeting. These farmers showed up and the conversations began.

That's another thing about this process: we're setting up spaces where creative problem-solving is possible, setting up tables where ownership is not by one party, but it's a space where people can come together and feel safe.

So we brought these farmers together, and they met with the chefs and the food buyers and the dining directors. They asked each other, "What would a relationship look like? How could this work?" Then they had organic college nights where the farmers came up, and the students loved it. That summer [they] set up a single-invoice, single-delivery system through ALBA Organics that allowed all the farmers to pool their resources, and allowed the institution the mobility (because they don't want twelve different trucks driving up two to three times a week with twelve different invoices), a system of compatibility, a relationship that could work on both ends. And suddenly we're launched the next year.

For the first year, '03-'04, there was only about $30,000 sold from this consortium. The following year, in '04-'05, there was, I think, around $87,000 sold. In '05-'06, it was $128,000. It's growing, which is wonderful.

This year they just signed a three-year contract with ALBA Organics and the consortium. We just had a meeting yesterday. All the farmers showed up, and we sat and talked about, how do we grow this relationship? What type of further educational partnerships are possible? Then we met with all the chefs and production managers from Dining Services, and they shared their excitement and enthusiasm and how to make this grow. The continuation of relationships, the potential for collaboration, and the ability to sit with one another makes all the difference.

In the summer of 2004, I was contacted by students at the University of California, Santa Barbara that wanted to [adopt UC Santa Cruz's food-sourcing model]. The University of Arizona, and the University of Hawaii at Honolulu, and several other institutions contacted Santa Cruz saying, "So you stepped away from Sodexo. How did you do it? Oh, it's working. How do we do it?" And in particular, in the UC system, [dining services at] eight out of ten campuses are independent/self-operated, and are interested in working with local growers.

Here at Santa Cruz we did develop some path-breaking relationships and were able to share those best practices. And we first convened a group of students, part of the statewide California Student Sustainability Coalition. Now this network, when I started working at the Student Environmental Center, had launched a campaign called UC Go Solar. Within a year and a half, the UC Regents passed the most broad-sweeping alternative energy and green building policy in the nation, that made UC the largest institutional source of alternative energy in the nation, catalyzed by students, partnering and trained by Greenpeace at the time. And following that, subsequently they are looking at a system-wide transportation policy that they're passing.

So we began conversing. What about food systems? How do we give the green light to campuses, give them the mobility they want, with the exercise of their own command, but with the influence of support from above? And we realized we needed a new component to this system-wide policy on the food system.

The Statewide UC Sustainable Food Systems Initiative campaign began at UC Santa Barbara, when we convened students from across the system in October 2004. Our goals were to be a network and clearinghouse of resources and best practices to mobilize the establishment of collaborative working groups, like our Campus Food Systems Working Groups; to provide networks and connections to local food producers and resources to make those connections possible; to establish campus, as well as system-wide guidelines of procurement, modeled after Santa Cruz's. We delivered over seven thousand postcards to a UC Regents meeting in 2005 at the University of California, San Diego. They had to carry those all the way back up to [the UC system's administrative offices in] Oakland. But we received endorsement from the UC Regents and the Office of the President to launch a system-wide task force for this policy, under the system-wide housing directors.

Suddenly we were being able to play in this bigger picture. At one of the most recent regents' meetings, at the University of California, Los Angeles, the regents said they were excited to see that policy come before them to vote on January 2009. So we are looking at a system-wide sustainable food policy that includes procurement, waste reduction, energy and water initiatives, as well as educational components and commitments, that hits a ten-campus system—the largest research university in the world—based upon the students' mobilization and the recognition of this need to change.

And we couldn't stop there. Across this time we've built great relationships with the state system, the community college system, and private schools across California. We also built [national] networks, as we're getting calls from students and groups across the country. I've traveled to the Northeast, up into Maine, to Florida Gulf Coast University in Florida, all around, sharing these practices, advising, and helping build technical infrastructures to make successes possible.

We began to talk about, what would it look like to create a national movement around this? The folks at the table for that included The Food Project's national program director, Anim Steel, who is a dear friend of mine; the Community Food Security Coalition's farm-to-college director,

Kristen Markley; the University of New Hampshire's Center for Sustainability, Tom Kelly; John Turenne, who is a twenty-five-year veteran of a [contract food service provider corporation], who now runs Sustainable Food Systems, Inc., a consultant firm that advises on institutional transformations of this nature; United Students for Fair Trade; the Coalition of Immokalee Workers; Student Farmworker Alliance. We brought in a cadre of people across the issues to say, "We need to focus on the big picture with purchasing. We can't just fight for fair trade, or local produce, or organic options, or humane food. We need it all. We need *real* food."

And hence this initiative called The Real Food Challenge was born. It's been growing in leaps and bounds. It's just so exciting. It's modeled after the Campus Climate Challenge. We're going to be challenging institutions of higher education to go from 2 to 20 percent sustainable food by 2020. That's a four-billion-dollar change of food sourcing, a new four-billion-dollar market to affect four thousand institutions, just to begin. It's a reachable, attainable, but challenging goal.

We're going to be assisting with training, and have regional and area organizers, and campus liaisons, and offer tool kits and resources and networking opportunities to make this possible. Our board of advisors, right now we have Anna Lappé, Michael Pollan, Vandana Shiva, and the list keeps growing. We're going to have networks, and organizers, and technical support from across the systems. We're working with all the software companies that supply major institutions with their software packages for their food designs to have technical assistance, Real Food Challenge packages. We're partnering with the National Association for the Advancement of Sustainability in Higher Education—the newest, fastest-growing, five-hundred-member institutional higher-ed association for sustainability—to incorporate our Real Food Challenge into their systemwide sustainability assessment package called STARS that they're beta testing this year on ninety campuses. We have a Real Food Challenge calculator that's a multivariate tool that lets people know where they're at, and where they can target and go.

It's such a tremendous opportunity to really roll this out, and Santa Cruz is at the center of it. We fire-started a national movement and recognized that there're chargers and champions and people ready to meet this challenge across the country. At the Center we've received the first national USDA Farm-to-Institution grant. Partnering with UC SAREP [Sustainable Agriculture Research and Education Program], the University of California, Davis, and CAFF [Community

Alliance with Family Farmers], we've interviewed over ninety-nine institutional food buyers, over fifty distributors across the state; did a national and state sample of students interested in sustainable food. And we've learned from all of these players what it takes to make it possible. We just got news that we have a special research grant through [U.S. Congressman] Sam Farr and the USDA that will help fund our program developments regionally, on the campus, and statewide for sustainable food in higher ed.

And from there we'll see. We want to roll out to the California State University system, the community college system, and learn about the different systems, as well as building bridges with the private campuses across California.

I had microwaveable food when I grew up. I ate fast food. Now I'm experiencing the most amazing food and seeing this transformation around food, realizing everyone eats, and the powerful transformative potential when people reflect on their food choices and look at how it can have a systemic effect. Changing our food system is going to change the world we live in. We're talking about developing solutions to some of the most current, cutting-edge issues and problems of our times and civilization. Youth can be involved in understanding their ability to transform their institutions and become civic engagers in changing our country and our culture.

I'm excited. I think that there's a lot happening, not just on the "blue" coastlines. This transformation is happening across this country. I've spoken with youth. I think that if we don't begin developing curriculum and education that matches the transformations necessary, we're going to lose the ability to be problem solvers in what we're inheriting in this third generation. It's not Generation X. It's the generation of inheriting the industrial revolution. It's inheriting the challenges of our times: the wars, the famines, the great inequities, the financial crises, and the environmental devastations that heretofore have not been unleashed—all in one generation.

We're carrying a burden that is immense, and we have children growing up not knowing what to do in this situation and tuning out and playing video games. We have parents advising them to do that because the world is too unsafe for them to go out and play in their backyards. We have a crisis. If this generation does not gain the respect and support of their elders, and is not given the tools to start making serious changes, we are in a loss of future and future and future generations, our grandchildren's grandchildren's grandchildren.

There's such a positive potential for people to connect, and I think food is an instrument to do that. I'm looking at dedicating my life to that. In just five years, I'm seeing transformations across the entire state and mobilizing national initiatives that have real substance. The energy is ready to make that happen, and we just need the response from other stakeholders to recognize this opportunity.

Paul Glowaski

Paul Glowaski was Garden Director for the Homeless Garden Project in Santa Cruz, California. The Homeless Garden Project runs a two-acre market farm that trains low-income and homeless community members in sustainable agriculture. They provide organic fruits and vegetables to Santa Cruz County residents through a community supported agriculture program. As farm manager of the Homeless Garden Project, Glowaski brought together his lifelong passions for economic, social, ecological, and food justice. Glowaski left the Homeless Garden Project in 2010 and is running Dinner Bell Farm (with his friends Cooper Funk and Molly Nakahara) on thirty acres of land nestled in the foothills of the Sierra Nevada Mountains.

I just had my thirtieth birthday. I grew up in Fort Wayne, [Indiana], and that's a city of about 300,000 people. But my grandfather owned his own farm. He had 600 acres of corn and soybeans and winter wheat, and he had a herd of 140 or 150 certified Black Angus cattle. I would spend my summers going up and being with him and felt a strong connection to him.

My grandpa died when I was a senior in high school. He died of pancreatic cancer. When he was farming in the seventies that's when you really started to see a lot more chemicals and pesticides used in agriculture. It was that Green Revolution. Former Secretary of Agriculture Earl Butz is from Lafayette in southern Indiana, Purdue University. Indiana in many ways was ground zero for this Green Revolution that was happening in the farms in the Midwest. The family farm started to change. Back then they didn't know how bad these chemicals were. They didn't talk about it as much. My mom says she remembers stories of my grandpa stirring the pesticides with his hands. And what's really terrible is that now that these pesticides are illegal here, you hear about farmers in Mexico or India, or you go there, and you see the same things happening. The farmers don't know. They're stirring the pesticides with their hands. Pancreatic cancer is very much an environmentally based cancer. There's no [absolute] proof, but you've got to think that farming like that probably wasn't the healthiest thing for my grandpa.

The other thing that I'll never forget is that my grandpa was just as willing to talk to the Amish folks as the immigrants in the area, to anybody. When he would drive around in his truck he would give this little wave to everybody who went by, this little flick of his fingers. He was Citizen of the Year. You hear this so much from old-timers when they talk about farming, that it's all about relationships with the people around you, and community, both the community of people but also the community of the land, too. Relationships are what my grandpa did best and what I admired the most [about him]. If he saw something interesting at someone's farm, he would just stop and start walking around. If someone came up and said, "Hey," he just was the friendliest guy and would introduce himself and become friends. I think about him often. I wear his jean jacket that my grandmother and mom gave me three years ago when they figured out that I really did want to be a farmer, that it wasn't just this phase of my life. It was going to become a way of life for me.

I went to DePauw University in Greencastle, Indiana. I studied Latin American history and spent a lot of my time with farmers in Mexico. I went on a delegation to Guerrero and Oaxaca and Chiapas. We went with some women whose fathers were farmers in Oaxaca and had been arrested. Their moms were living in the Zócalo in Oaxaca City, trying to bring awareness to their situation. We went with these women to Guerrero and to Chiapas so they could communicate their story to the different indigenous groups who were all feeling the same effects of neo-liberalism in their very remote communities—desertification and highways being built and a lot of violence. We were part of the Chiapas Media Project in the Mexican Solidarity Network, trying to help bring communication. We didn't go there in the spirit of "we need to do something for these people; we have something to teach them," because really, they had everything to teach us.

We stayed in an Aguas Calientes, one of their autonomous zones that they had set up in Chiapas. I stayed with this farmer. We went out and harvested *frijoles* out in the field. You hike two miles and you get to these beautiful bean fields up in the mountains. We would harvest into baskets, and then put them into burlap sacks. After my grandpa had died, this was the first time I had been involved in agriculture again. I remember they had straps around your head, and you would carry the beans and hike the sacks down through the mountains back to the Aguas Calientes. They'd put them out on a concrete pad to thresh and winnow. I remember we were about to leave, and I was feeling like, what the heck am I supposed to do now? I have nothing to do for these people. I want to do something. So I asked this farmer, "What can I do for you?" He said, "Only when there's justice where you're from will there be justice where I'm from." This farmer out in the jungle ended up probably saying the most profound thing that anyone's ever said to me in my entire life.

I ended up joining AmeriCorps when I graduated from college. It was teams of young people based out of Charleston, South Carolina. We would go and do projects around the Southeast. We worked in a school, and we did Habitat for Humanity, and went to a swamp in Georgia. I taught kids how to ride horses in Tennessee. I'd never ridden a horse in my life. We did disaster relief in New York City after 9/11.

This desire to serve definitely came from my mom and dad. My mom has spent her whole life as a social worker and caregiver, taking care of women after they've had abortions, or working with kids who've been abused, or helping people who are sick and dying, families after a family member has died. She's this really loving woman who is very much interested in serving the people around her in her community.

And my dad, he has had Crohn's disease over thirty years, since before I was born. He's this super stoic, courageous guy. He was laid off from his job this past year, of twenty-nine years. And even when he lost his job, trying to find something to do, he went and he started to tutor kids. He's tough as nails. In Catholicism right before you die, they give you anointing of the sick, like right before you're going to die. My dad's had that, like, five times. So he's a real tough guy. He worked at [a moving company] in a corporate environment. He worked at that job so my brother and I could go to college, and so he could provide for us. They laid him off after twenty-nine years. He would have been able to retire in three months. It was criminal, evil kind of stuff. Now he runs a little food bank at a church. He essentially runs a low-income community supported agriculture project. My parents have been heroes for me as far as shaping what I wanted to do and where I would go.

I applied for another year in Americorps. I moved to Vermont and worked at the Committee on Temporary Shelter in Burlington, Vermont. My uncle is a homeless Vietnam veteran from Fort Wayne or from northern Indiana—my mom's brother, my Uncle Mike. I remember being in high school, driving in Fort Wayne and seeing Mike walking around Fort Wayne in the snow and picking him up and giving him a ride. He definitely has some demons—alcohol, to be sure. Even our family member was just devastated, and we couldn't even help him. I think that's probably why I chose to do AmeriCorps at a homeless shelter. I was hired on for another couple of years and stayed and did community outreach with homeless folks in Burlington.

I went to the Republican National Convention in New York City, in 2003, to protest. I had been working with homeless folks and homeless vets. My friend's band was playing in Brooklyn the night of the big marches and protests. They were playing in Brooklyn on a rooftop overlooking the city. It was this beautiful view. There was posters everywhere, and everyone's wearing political T-shirts, and it was very inspired. We were there to party and be with a community of like-minded people. People were feeling like we've gotta do something here now. There was a lot of energy, and it felt very exciting. Most of the people there were people my age, white, kind of the standard crowd you'd see at a rock concert, and at a rooftop in Brooklyn, but there was an older African-American guy there. I went up to talk to him. I'm feeling real excited and empowered. I said, "Hey, my name is Paul. What's your name?" He told me his name. I said, "So where are you from?" He said, "I'm from Chicago." And I said, "Oh, cool, I'm from by Chicago. I'm from Fort Wayne." So we started talking about Chicago and Indiana. I said, "So what do you do?" He goes, "Oh, well, I used to be homeless, but now I'm a beekeeper." I said, "You're a beekeeper?" He said, "Yeah, I keep bees in the city. Those kids over there," and he pointed to some young white, mid-twenty-something kids, and he said, "They taught me how to be a beekeeper. Now I'm doing pretty good."

I thought, oh, my God! The whole world opened up to me. To have bees in Chicago? And this guy was homeless! This was another one of those times where it just really—that man changed my life, because I knew what I could do. I could teach people how to grow food and to farm and that we could get self-sufficiency and food into our communities that way. Working in the homeless shelter was just Band-Aiding the problem. Homeless shelters just mask what's really going on. I remember thinking, maybe we should just close all the homeless shelters for a week so people actually have to see how many people are homeless, and see what that's like.

[I came back and started] to research urban farming on the Internet. Urban rooftop farming was what I was going to be into. I was going to farm on warehouses in Chicago. I had it all figured out. These buildings could withstand the wind in Chicago. They're built very strong because of the wind. These buildings can withstand huge loads of soil and snow on top. I remember reading a paper about urban agriculture projects around the country, and one of them that I read about was the Homeless Garden Project. I never in a million years thought I'd end up in Santa Cruz, working at the Homeless Garden Project. Even then, when I heard about it, I thought, well, that's pretty cool! That's a great thing, but I'm going to move to Kentucky to learn how to farm.

I knew nothing about farming, really nothing. I didn't know what to do other than to go and work for someone. And my friend Jake Schmitz, who works for Organic Valley now—he was from Kentucky and said, "Oh, I've got these great farmers you should move to Kentucky with." He called them the Mickey Mantles of sustainable agriculture. So I left Vermont and my community and all my buddies and decided, I'm going to move to southern Kentucky, down by Bowling Green, to learn how to farm so I could move to Chicago and do this rooftop farming gig. Paul and Alison Wiediger owned and still own Au Naturel Farm in Edmonson County, Kentucky. Smiths Grove is the name of the town. Edmonson County has no stoplights in the whole county and is a dry county. There's no booze. It's a very southern, rural sort of place. I went there to learn from them because they were using these high tunnels, these unheated greenhouses, and they were able to grow food twelve months out of the year, even though there was snow on the ground, which I thought was pretty cool stuff.

[Later] I ended up at a farm [of] this older tobacco farmer in Kentucky. They early on knew that tobacco's time was coming. It was pretty savvy. So they used all of this government money that was being doled out to help farmers convert and created probably the largest or one of the largest organic farms in Kentucky, a diversified market farm, but way bigger than Paul and Allison Wiediger's family farm. This was a giant, 300-acre thing, where they were selling at farmers' markets four or five days a week. I spent that summer helping at the farmers' markets and helping to set up for the CSA each week, in their first year of a CSA, just trying to figure that out, and then being the assistant for the pastured poultry piece, taking care of 2,000 broiler chickens and the few hundred egg layers over the season. We had 150 turkeys that I cared for every day. I worked with them for the year, and it was hard.

Then I applied to the Farm and Garden at UC Santa Cruz. I got into the apprenticeship and drove cross-country to California to go to CASFS [The Center for Agroecology & Sustainable Food Systems]. That was amazing, being at the Farm and Garden, for so many reasons, to immerse yourself into that type of environment where everyone is positive and excited and engaged. Really beyond the friendships that came out of the apprenticeship, just that idea of helping me to ask the right questions was so powerful. Now, when I walk onto a farm after being at CASFS at the Farm and Garden, I can almost immediately see what's different. I can see that a cultivator is set up a little bit differently, or maybe their bed shape is different, or maybe they're overhead-watering some dahlias. I think, at CASFS we would never do that.

They give you the base and the tools to be able to farm. People like Orin Martin, and Christof Bernau, and Jim Leap are such powerful teachers, heroes.[1] You get affirmation. Whenever you got any affirmation from Orin, oh, this is the most amazing thing. If he would tell you, "You did a good job on pruning an apple tree," you know, that was just the best, because everyone looks up to Orin. He has impacted so many people in his time.

But ultimately, it was the peer support that ended up pushing me forward and making me believe that I could do whatever I wanted to do. It was really my peers, talking about agriculture twenty-four hours a day, staying up late talking about food security with people and why it was important that we all had good food, and the speakers that would come. Vandana Shiva came and ate lunch with us, or dinner. I knew really very little about this woman at the time. Now [I] have read a lot of her stuff. She comes on the farm, and she's very soft-spoken, an incredible listener, a very quiet woman. After dinner, we went to watch her speak up at the university, and she was railing against chemical companies and was so loud and was a strong woman pounding her fist against the podium saying, "This is wrong, what's happening to farmers and Indians." Meeting people like that and going up to People's Grocery, and when Bob Scowcroft[2] came and spoke to us, and Joy Moore, who

---

[1] See the oral histories with Orin Martin and Jim Leap in this book. Christof Bernau is the garden manager at the Center for Agroecology and Sustainable Food Systems.
[2] See the oral history with Bob Scowcroft in this book.

started Fresh Farmer Choice up in Oakland—hearing people like this. I was so excited when I left.

Of course I'm in Santa Cruz, and there's the Homeless Garden Project right down the road, so that was one of the first places I went to when I came here. I thought, I can't believe I'm here, after hearing about it three years earlier. While I was an apprentice, I went and I taught a class on these high tunnels that I learned about from Paul and Allison. I would go and do work days at the Homeless Garden, and I was the apprentice liaison to the Homeless Garden.

I started [my full-time job at the Homeless Garden Project] April 12, 2007. We really had to hit the ground running. They had had a tough year in the greenhouse, a tough winter, and they didn't have many starts. We were not really in good shape, and our CSA was supposed to be thirty-nine members that year.

At the Homeless Garden Project our farm was set up in this French-intensive, biointensive, labor-intensive type farm, the four-foot-wide beds, high levels of fertility, the same way that the down garden and the up garden [at UC Santa Cruz] are set up. And really that's because Patrick [Williams] had studied with [Alan] Chadwick up in Covelo.[3] Patrick is another one of these Chadwickians that are running around, changing the world by starting gardens and farms. He's gone all over the world, helping to start bio-intensive type farms and gardens and had landed at the Homeless Garden.

The Homeless Garden was started by Dr. Paul Lee,[4] who helped, along with Page Smith,[5] to bring Chadwick to the Farm and Garden. Page Smith and Paul Lee were professors. At any rate, they were the ones who brought Chadwick over. And it's interesting because the transitional shelter downtown is named the Page Smith House, and there's now the Paul Lee Annex. These two people who helped change agriculture in general were also very much influenced by homelessness. The reason the transitional home is named after Page Smith is because in the eighties Page Smith helped to start the first homeless shelter in Santa Cruz.

---

[3] In 1973, Alan Chadwick started the Covelo Village Garden Project in Round Valley, California.

[4] See the oral history with Paul Lee in Maya Hegege and Randall Jarrell, *The Early History of UCSC's Farm and Garden Project*, (Regional History Project, University Library, UCSC, 2003) http://library.ucsc.edu/reg-hist/farmgarden.html

[5] See the oral history with Page Smith, Randall Jarrell, ed. *Page Smith: Founding Cowell College and UCSC, 1964-1973*, (Regional History Project, University Library, UCSC, 1996.) http://library.ucsc.edu/reg-hist/smith.html

He was part of this group of community members who saw a need for homeless services in Santa Cruz.

There always have been those people kind of riding on the rails—the guys with the sticks and the handkerchief, some folks who are homeless. But these homeless people on the streets, that was something new. That hadn't been there before. Homelessness is very much a product of my generation. I was born in 1979. Today there's three million homeless people. There're a million homeless kids who leave school every day and go and live with their families in their cars. [Before] even if people didn't have money, people had a place to live. Shelters mask homelessness, but people living in their vehicles also mask homelessness. A huge population of people live out of their vehicles.

They're dealing with a lot of barriers. Once someone grows up in abuse, a lot of times they can either be abusive or just live in abuse. The cycle just continues. The same thing is true with homelessness. Deinstitutionalization, the folks coming back from Vietnam, people's wages not going up relative to the cost of housing in our communities—all of these things came together in this perfect storm.

I'm always trying to make sure that people realize the relevance of the Homeless Garden Project. It's been on the cutting edge of sustainable agriculture for almost twenty years now. Well before I ever had the idea about the importance of food and nutrition in people's lives, some folks in Santa Cruz realized it, when I was still a young kid. They set up the Homeless Garden Project.

We're working with thousands of volunteers each year in addition to our trainees. On any given day at the Homeless Garden Project, you'll have the richest to the poorest in Santa Cruz working together, farming together. And we eat together every single day. We sit down over a meal. On the best days the homeless are teaching the housed how to grow their own food, a real role reversal.

We've been around for twenty years. I think that is our blessing and that's our curse. At times the Homeless Garden Project can be old news to people. It's not quite as sexy and young and hip. But it *is*. It is. These ideas about the green economy and local food—the Homeless Garden Project is part of that. You hear Van Jones talking about green apartheid. We've been talking about green apartheid for twenty years, that the green economy can't just be for rich people, that the green economy has to be for all people. We've been training homeless folks for so long. Over five hundred people have gone through the training program at the Homeless Garden since its inception. Tons of people have had their first experiences farming at the Homeless Garden Project, their first opportunity to be involved with agriculture. We have a lot to teach people. We have staying power. Today, more than ever, the Homeless Garden Project is needed.

Steve Kaffka

Steve Kaffka (left).

Stephen (Steve) Kaffka came to UC Santa Cruz as a philosophy student in 1967 and began volunteering in Alan Chadwick's Student Garden Project. He worked side-by-side with Alan Chadwick and became the student president of the Garden in 1968. After Chadwick left, Kaffka managed the Farm and Garden and formalized the apprentice program through UC Santa Cruz Extension. Kaffka has achieved a distinguished career as a research agronomist. After he left UC Santa Cruz in 1977, he earned his Ph.D. in agronomy from Cornell University and now directs the California Biomass Collaborative. In May 2008, he was the subject of a National Public Radio documentary, "Are Organic Tomatoes Better?" which featured his research comparing the nutritious value of organic versus conventionally grown tomatoes.

At UC Santa Cruz I majored in philosophy. I had gotten swept up in the anti-Vietnam War mood of the time and was pondering both my own relationship to the larger society and also potential participation in that war when I would graduate. I decided that that war was a manifestation of the larger cultural and social forces operating, directing American life, and that if I didn't want to have to be a part of that, or wanted to somehow find an alternative way to live, I needed to learn how to grow my own food.

So I went up to the Student Garden Project in the spring and started working there. I figured out that I could, if I lived in poverty, work there for the summer, because there was food at the Garden, and you could sleep in the woods if necessary. I worked there full time that summer, and I took to it. I got to know Alan Chadwick and ended up being named the student president of the Garden in fall 1968. There wasn't much of an organization, really. We didn't have a treasurer or vice president. Essentially, I was nominated to be president by Alan, which was a British aristocrat's notion of democracy.

I'm not sure why Alan chose that spot for the Garden, but sometime later, I saw a picture of an older hillside garden in Britain with beds running up and down a hillside that looked almost identical to what he created. It had been from Britain somewhere on a south-facing hillside. He may have seen something like that elsewhere.

I would often give visitors tours of the Garden. Grant McConnell, who was a professor of political science in the early years in Santa Cruz and had been very active in the conservation movement, was teaching at Santa Cruz. He and his wife, Jane McConnell, were interested in the Garden. Grant and Jane had been very active in helping establish the North Cascades National Park, with the help of David Brower, who was at one time the head of the Sierra Club and then Friends of the Earth. So he brought David Brower to the Garden. I showed them around. I started the tour off saying, "The Garden at one time looked just like this chaparral-covered hillside. Now look at how wonderful it is." Grant told me later that David said to him, in an aside, that the chaparral-covered hillside looked just fine to him. I've always remembered that, because in that anecdote you have a juxtaposition of two views of nature and what humankind's relationship to nature should be. Brower is concerned with preserving nature, while Alan saw the original hillside as a place to transform and create. I see these contrasting views in all kinds of places—in the Upper Klamath Basin, where I have worked, and other places. You have people who have an ethos that says: work, change and transform nature, make something from natural resources. They're farmers, or they are people who have natural resource-based livelihoods. Then you have others who say that the best way to treat these resources is to step away from them and leave them alone.

We would invite selected faculty up to the Garden for lunches. It was quite a nice event—you know, flowers, and on the porch of the chalet, vegetables from the place. Alan was a first-rate cook. One of the little things that I got from him was an appreciation of what good food is, really good food. The vegetables that we grew, he used, certainly, but he would buy other things. We would often grill salmon from the [Santa Cruz] Wharf for people. Alan was a real sugar addict. He would make these tarts that were essentially a big can of jam, with a killer crust baked on top of the jam, for tea sometimes, bring them over, and then he'd put sugar on that. And espresso coffee at four in the afternoon. I've never had better crusts.

Giving away flowers was [another] thing that made people love the Garden. There was a little bus kiosk across from the Garden that was adopted as a flower stand. People would stop in the turnout and go across and get flowers. There were complaints from some of the florists in town, because there were thousands of flowers given away every day. If you went in any office around the campus, people had them. It was such an extraordinary idea, to give them away. But he would get mad at people because they would knock flowers out of the water. There was always that contrast between this fabulous generosity and then the fuming and anger about it when people didn't conform to his sense of rightness.

The largest event that I can remember during that time—I think it was in about 1970, in the spring. A man named Francis Edmonds came to visit the Garden. Francis Edmonds was the head of a small school in England called Emerson College, [which] was focused on the teachings of Rudolf Steiner. He came because my roommate, and still good friend, Jim [Pewtherer], had made some contact and was getting interested in the Anthroposophical Society. We had salmon for 150, a sit-down dinner at the Garden. I grilled the salmon. I had the good fortune of being at the table with Francis Edmonds. Alan had not talked very much about Rudolf Steiner and anthroposophy in those early years, and he never used the term, in my memory, "biodynamic." At one time, people would refer to the raised-bed, intensive gardening that he taught and introduced to the U.S. actually, as French intensive gardening. That's what Alan called it. Later on, after Edmonds' visit, he would begin to talk more about Steiner, or at least anthroposophical notions, esoteric notions. The term "French intensive biodynamic gardening" started to be used, but it wasn't really part of what he talked about initially. Alan never used what are called biodynamic preparations, ever, and I don't think he ever did in any of his gardens. I think Alan was one of the more original interpreters of Steiner's ideas, not simply a rote follower of Steiner's suggestions or notions.

Alan started to become a bit more involved with the anthroposophical circles in California. A group of us went down [to Southern

California] with him that summer, and he gave lectures at Highland Hall, which is a Waldorf school in North Hollywood. Everybody had dinner at my parents' house. Alan was always very dramatic and a wonderful person to listen to, no matter what he said. People in anthroposophical circles were interested, so he started to do a bit more lecturing. From that time on, he started to talk about biodynamics as a more generic version or term to encompass some of the vitalist ideas that he had about gardening and nature.

When Francis Edmonds came [to the Garden], he suggested [that] he send Herbert Koepf to Santa Cruz, [and that] Herbert Koepf would like to see this place. If you were to read about biodynamic agriculture, you would find his name as the principal name in the late twentieth century. He wrote the standard book on biodynamic agriculture, in German and in English, and was in what they called the Vorstand [management board] in the Anthroposophical Society.[1] There were seven leaders of different divisions of the Anthroposophical Society, and Herbert was the leader of the biodynamic agriculture movement worldwide and was also teaching biodynamic agriculture at Emerson College. So the next summer he came, and he stayed in my house on Western Drive, which was short on amenities and very primitive, really. It was really extraordinary. Here was another middle-aged man who was willing to stretch his limits. Herbert came and gave a lecture about biodynamic agriculture in its more formal, Steinerian sense.

Alan didn't want to have anything to do with him. He was almost rude. Herbert was there, but Alan didn't interact with him at all. It surprised me. I think partly Alan's reaction may have been a residue of Herbert being German, but he was also a separate authority, a different authority, different from Alan. Whatever the reasons, they didn't interact at all. Herbert subsequently returned several times after the Farm had started, at my invitation. He came on two or three occasions to give lectures and again stayed with us. He was a soil scientist and had been in a German university professorship in soil science at Hohenheim Univeristy in Stuttgart. But he also had the capacity to tolerate cognitive dissonance, to handle, or be interested in and committed to, alternative agriculture notions. I hadn't met anybody like him before. He tried to integrate those different ideas. He became something of a mentor for me. Subsequently, when I left Santa Cruz and had a Fulbright to go

---

[1] The Anthroposophical Society was founded by Rudolf Steiner in Switzerland in 1923.

to Germany, he helped set that up. We wrote a couple of papers and a book together, later. So it was a lifelong, enriching connection for me.

When I graduated in 1970, I stayed on to work full time in the Garden. When Alan would go off to give talks, I would always go with him. He'd ask me to come with him and help him drive, and I went to some very interesting events. I had a fairly close relationship with him, and as the president of the student organization helped get funding for the Farm.

I was graduating and the Vietnam War was raging, [and] with Alan's encouragement, I applied for conscientious objector status. I was fortunate enough to actually get it. UC Santa Cruz Chancellor Dean McHenry wrote a letter supporting me, even though he said he didn't believe in that view. But he believed that I was being honest and truthful. I got CO status, and I was assigned to work in the Garden. That was my alternative service, instead of a hospital or whatever. I had said, "We're building this farm, and we're creating alternatives to the kinds of social institutions that lead to conflict and violence." Perhaps the chancellor's letter mentioned that, I don't know. I apparently had a reasonable narrative for them that they believed. So from that summer on, that was part of my reason for being there. I was really very committed to those ideas; otherwise I don't think I could have been persuasive to the draft board. I certainly was showing up six or seven days a week.

The Garden reached its height in 1970, '71. It was bursting at the seams. It was always called by Alan "The Student Garden Project," and "The Student Farm and Garden." That was the whole rationale for having it there, so that students could learn. The problem, of course, was that the labor was too much for students. Student participation, at best, was always inconsistent, including my own, because of conflicts with studying. And the levels of skills that Alan demanded were greater than people walking up the hill could ever have to start with, because no one had been trained in that landscape gardening tradition. Nobody ever. So there was always a need for people on a full-time, regular basis. There was always, always this problem of how to reconcile the labor needs, and the quality of the labor required, with random student participation. That was only finally solved when we started the apprentice program.

Tensions started to grow fairly significantly around his style of overreacting. Alan was always ambivalent about the university. He periodically would suggest that everybody should pack up and leave with him and go to various places that he had in mind. The first place was New Zealand. He wanted whatever group of students and people who were working at the time to just go off and form a little colony in New Zealand. Then he got interested in the Seychelles Islands. He was very serious about it. He described the place and how the university wasn't the right place for what he wanted to do. The university was crazy, and there were all these terrible ideas about biology and science, and it would be far better if we all started off in a new place.

The idea Alan had about a master-apprentice relationship, that he would be the only source of information or authority in a place where ideas and broad learning are, in fact, the currency, there was an inherent difficulty about that. I mean, people like [biology professor and founding Crown College provost] Kenneth Thimann really thought that he was just full of nonsense—that he was crazy, that he was teaching erroneous stuff—and was not very happy about Alan's presence, or at least what he was teaching. I'm sure that must have been evident to Alan, and he probably felt like a fish out of water there, really, in the end, in terms of what he had thought was the best way to do things, and how to live.

Alan Chadwick was anti-science. He talked a lot about how terrible it was to do reductionist work, and how, if you break a thing into its parts, you could never understand it as a whole. It's quite interesting, because in the last few years I came across a [1940] book called *Look to the Land*. It was written by a man named Lord Northbourne, who was a British squire. The organic movement in Britain was actually started [at the] beginning of the twentieth century and in the late nineteenth century. It was really a reaction against modern developments, not only in the landscape, but also in social structure in Britain. It was largely driven by organizations that were Christian in their character, and saw life on the land in the peasant-squire style, and maintenance of soil quality and recycling of manure, and mixed farming of livestock and cropping, and all of those things as part of a whole, a whole cloth. If you read that book, *Look to the Land*, you can hear the ideas that Alan had about the relationship between people and nature. They're all in there, not in quite the same words, but essentially. Alan's views about how agriculture should be organized, and what it was, come out of that historical context. It was quite striking for me to read that book and have it echo in that way. Northbourne was the first person to use the term "organic farming," in English, anyway.

The work ethic that Alan brought and embodied was basically the only thing I knew as well: there was only one way to do things, and that was to work constantly, dawn to dusk, which everybody did. Actually, when you're young, it's a glorious thing to do. It's a glorious thing to be so in your body. I think one of the most powerful things that Alan taught had nothing to do with particular gardening techniques. It had to do with being in your body, and being able to physically effect change in the landscape with your muscles and have [a] direct, intimate connection with the land.

After the first summer in the Garden when Alan used a rototiller, before I was there, he got on an anti-machinery bent, and there were no machines ever again used in the Garden. The idea was transferred down to the new student farm [after Chadwick left UC Santa Cruz]. We were going do everything, as much as possible, by hand. But because there were large areas that had to be cultivated or could be cultivated, we actually were interested in using horses. Again, I just don't see how this was even possible to have happen at the University of California, but we bought a team of horses. Nobody knew how to use them. We bought it from this eccentric guy named Jim Baccus. They were unmatched. One of them was crazy. They were draft horses. They were big animals. We assembled a bunch of old equipment, started to figure out how to use it. But the horses didn't work out very well, not surprisingly, because nobody had any skill. There was a lot of old, rusty stuff still around [from the historic Cowell Ranch] and especially forty years ago you could still find little [horse-drawn] disks and stuff. One day the disk was being used in among the apple trees with the team, and because the person driving the horses wasn't skilled (because no one was skilled), they knocked a sprinkler riser head off and caused a gusher, and that caused those horses to bolt and run. The driver fell off, landed on his rear end on the disk, which sliced him to the bone. He was fine, after some stitches. There was no permanent damage.

I had in mind always that the UC Santa Cruz Farm and Garden would be university resources for the broadest possible community, but particularly for students. I knew that I would never be able to be charismatic and inspiring like Alan, but I had hoped that [after he left] at a minimum we could teach some of what he taught, and that people would continue to have the opportunity to experience what I had experienced, which I had found so powerful and life-changing, to be able to work in this way, physically and in a holistic manner with growing things. I found that experience transformative. I feel largely vindicated in that, because the fortieth reunion event a couple of weeks ago was a demonstration that people still feel that way about the place. That feeling, that transformative experience is still accessible there. It's become a culture in the place that's not dependent on anybody anymore, not on any individual. It's literally a culture. So you can think of this place as having given birth to an element of culture. It's the genesis of a movement. The Garden was right between Big Sur and San Francisco on the coast. It was organic from the beginning. It was fabulous. It was at the right place at the right time. It was right in the middle of those two centers of the back-to-the-land movement and the alternative culture of the late sixties and early seventies. It was centrally placed and timely.

Richard Merrill

Richard Merrill is perhaps best known for editing the 1976 anthology Radical Agriculture, *a formative text in the sustainable agriculture movement, along with the 1978* Energy Primer: Solar, Water, Wind, and Biofuels. *Merrill helped start the El Mirasol urban organic farm in Santa Barbara, California. In 1975, he founded the Environmental Horticulture Department at Cabrillo Community College, which he directed until retiring in 2005. While at Cabrillo, Merrill mentored and inspired several generations of students, who went on to become organic farmers, gardeners, and activists in the Central Coast region and beyond. Currently Merrill runs his own environmental consulting service, Merrill Associates. He recently coauthored with Joe Ortiz* The Gardener's Table: A Guide to Natural Vegetable Growing and Cooking.

I was born in San Mateo, California, in August 1941. I'm a third-generation Californian, and the philosophy of our family is, why leave this place? Here in California, we have seasons in space instead of time; you can always find a season in California.

I went to UC [University of California] Berkeley on a football and track scholarship in 1959. I blew out my knee and was getting bored with the fraternity life. This was one year before the Free Speech Movement hit the fan there. Anyway, I transferred to UC Los Angeles for eight years and got immersed in academia, studying philosophy, mathematics, and biology. I finally settled in biology, mainly because it's one of the few things I got all A's in. I needed a job. I ran across a man named Monte Lloyd, who was a professor of ecology. Now, you have to understand, this is before ecology became popular. This was in the late sixties, before the environmental-cultural revolution of the early seventies. Monte Lloyd was a true genius in the intellectual sense. He was the smartest person I ever met. He was interested in spatial arrangements of animals, and mathematics, and how you can look at the spatial arrangement of living things and determine a lot about how they operate. So I got involved in statistical biology.

He went on to Chicago, and his good friend, Joe Connell, who was probably one of the three or four most famous ecologists on the planet, was up at UC [University of California] Santa Barbara studying diversity and stability in ecosystems. I went off there. Joe was an absolutely amazing human being, one of my very favorite people. He taught me how to go out into nature and see things that other people don't see. "Not things," he said, "but patterns. Don't look for things, look for patterns. Watch how animals interact, not how they are. Take the time to watch how they behave." He would have me look at a flower for an hour. Just stare at it. He said, "Don't get Zen-y on me. Just look at a flower for an hour." And he was right. You become in to nature. There's no nontrivial way to say it. You just become part of nature, and then nature starts doing its dance around you. Then you become the observer that all these great scientists are that know how to look at nature and see the patterns. He taught me that.

Then, unfortunately, the riots in Isla Vista started. The Vietnam War was consuming me. I couldn't do research any more. I was protesting all the time, constantly, because the morality of the war was just unfathomable. It was unbelievable. I said, I've got to do something more relevant than this. I quit. I look back on it and I quit an absolutely amazing opportunity. But where I was going was research, and I didn't want to do research. I wanted to teach. People say, well, you could have done both. I say, no, not really. That's not the way it works in higher education. I went and helped start the El Mirasol garden in Santa Barbara and then just went from there. Never looked back. Didn't regret it ever.

I realized my politic wasn't Left and Right, it was up and down. It was centralist-decentralist. I hooked up with Murray [Bookchin], who lit me up with his very famous essay called "Ecology and Revolutionary Thought," which changed everything. Lots of people were changed forever by that essay.[1]

---

[1] "Ecology and Revolutionary Thought" was first published by Murray Bookchin under the pseudonym Lewis Heber in Bookchin's newsletter, *Comment*, in 1964, and republished in the British journal *Anarchy* in 1965. It synthesized ecology and anarchism and became a founding document for the emerging field of social ecology.

Then he came out to Santa Barbara, and the rest is history. He convinced me that I was not crazy because I did not think like everybody. Because to me, the problem was centralism. He was one of those people in your life that throws you a lifesaver, and opens the gate, and you're forever changed.

When I left Santa Barbara, I took that whole theoretical ecology, diversity, stability thing I had been working [on with Monte Lloyd at UC Santa Barbara] and [applied it to] agriculture, [to the] energetics of agriculture and the ecological efficiency of agriculture. The ecological efficiency, in the long run, is the same as the energetic efficiency. How you tie all that together was what I was consumed with most of my adult career. One of the most important concepts that came out of the seventies environmental movement was the "bottom line," the nature of net energy. We did our analysis with agriculture, but you can do it with anything. For example, you want to bring in soft phosphate from Florida. Okay, you mine it, package it, and ship that from Florida to California. Calculate the BTUs. Now make chemical fertilizer locally and I can show you that energetically it's cheaper to use local chemical fertilizer than distant natural ones. So if you are an organic agriculturalist, you are faced with a dilemma. Because the bottom line is not fossil fuels. The bottom line is energy efficiency. The monetary thing for tomorrow is energy credits, carbon credits—call them what you like. It's all the same thing. I've been discussing this with people in sustainability for decades. You can't just focus on natural this and natural that. You gotta focus on what's energetically efficient. The real problem isn't so much the net energy drain, but the fact that the current energy source, fossil fuels, is nonrenewable. Imagine a sensible civilization creating an infrastructure for making a renewable resource, plants, using a nonrenewable one, fossil fuels. It's quite insane, even from a clinical point of view.

I began networking with people throughout the country who were writer/activists in the area of alternative agriculture. It was a small community of like-minded people who were approaching change from a wide variety of directions: economically, ecologically, culturally, etcetera. It was obvious that no one person could accurately describe the alternative agriculture movement. I was gathering essays for a book. My good friend Murray Bookchin had written some political books for Harper and Row. He wrote to his editor and said, "You should take a look at these essays. I think this guy has got something." At the time it was called some ridiculous title. And he [Bookchin] looked at me right in the eye, and I'll never forget this, he said, "No." He says, "You're just denying this whole thing. You've got to call it *Radical Agriculture*. Go to the root cause of the problem, radical. This is not a liberal solution. This is a fundamental problem here." I said, "Okay, Murray. Okay."

So I decided to do *Radical Agriculture* as a series of essays written by experts in the various areas of activism. I found out that the whole alternative agriculture subculture, if you will, actually consisted of an incredibly disparate group of people who were coming at change in agriculture from all sorts of different ways. So the book wound up as an amazing, eclectic group of people. Murray Bookchin kicked it off with trying to explain what radical meant in that sense.

And you have to remember that in 1973 and 1974, when I was researching this, these people weren't so famous. They've become famous. This is an important point. Because when you look at them now you go, wow. Wendell Berry, of course, is the poet laureate of alternative agriculture. Wendell was the most famous one. He was pretty well known. Peter Barnes, I have no idea what happened to him. Or Nick Kotz, Sheldon Greene. Mike Perelman was a professor of agricultural economics up at Chico [California State University at Chico]. I haven't seen him for a while. He got me going on the whole energetics thing and was one of the more important people in my life. Jim Hightower, at the time, had written *Hard Tomatoes, Hard Times*, which was a very interesting book, an indictment of the USDA, but he was certainly no secretary of agriculture for Texas, which he became later. Then he became very famous. If I were president, he'd be my choice for secretary of agriculture without even batting an eye. He's, of course, extremely well-known now, and has his radio show.

I tried to get César Chávez to write a chapter. I had a conversation with him once, very brief. I asked him to give me some sense of the overall picture. Chávez said, "Well, I'll tell you something. Most of the strikes that we do have less to do with wages than with pesticides and reentry. They get couched and reinvented in an economic sense. But a lot of it, we're out there battling for the lives of our people who are forced to go in before the reentry levels of the pesticides are up." The original struggle started as a pesticide issue and not a money issue.

You can [look at the history of organic farming as starting] in 1924 with [Rudolf] Steiner.[2] Steiner gets confronted by a bunch of farmers

---

[2] The development of biodynamic agriculture began in 1924 with a series of eight lectures on agriculture given by Rudolf Steiner in Germany.

from Europe. They came to him because they had a problem: they couldn't get their seeds to sprout. In those days, you didn't buy seeds, you saved seeds, and they weren't sprouting anymore. The farmers believed it had something to do with the soil. So he laid down the basics of biodynamic agriculture in 1924, which has some rather esoteric stuff like putting aged manure in a ram's horn, but there's also a lot to it. It turns out that there is something to the moon planting. It has to do with gravitational pulls. But it could be as simple as moonlight. And there is something to the benefits of silicon in the soil. Plants don't use silicon, but it reflects light and it increases the photosynthesis of algae in the soil, which increases the microbial productivity of the soil. And although "companion planting" has long been replaced by a far more sophisticated palette of plant combinations and micro-habitats, the idea was an early dead-on. There's something to all this. But for me, the various natural methods were never connected as a whole biodynamics. To me the dynamics and challenges of a garden or landscape were crystal clear only when I viewed them as an ecosystem. It's just the ram's horn is not something I have figured out yet.

So then the other wing was organic, and that came about from Rodale, of course, in 1941. And he linked up with Albert Howard. Howard had been sent to India as an agricultural adviser. While there he developed a technique of rapid organic decay, composting, designed to relieve dwindling soil in the Indore Province of India. It came to be known as the "Indore" method.

The whole organic movement started with flower and vegetable growing. Vegetables are interesting because you take any vegetable: it started out as a weed, became an herb, and slowly they bred out the toxic chemicals that were unpalatable, and it became a vegetable. Virtually every vegetable went through that sequence. Lettuce used to be a weed. Then it was an herb, and in Turkey it's still an herb. So you go through this whole succession. The point of that is that vegetables are unique plants. They need an enormous amount of attention. In fact, if you want to know if you have a rich soil, note if the predominant perennial weeds are edible. If they come from edible stock, like dandelions, then it's a fertile soil. All vegetables were once weeds that were predisposed to rich soils. So you have to keep feeding them. I say, "You want a vegetable garden? You've now got a pet." They give you a lot, but you can't leave them alone. They're not cactus.

So the point of this is that Rodale started this whole idea of organic, compost-rich soil, so that people see organic as a rich soil thing. The truth is that outside of vegetables, annuals, a few herbs, and a few other subtropical plants, the vast majority of landscape plants thrive in normal, even poorer soil, as long as it's well drained. And so you have to learn that the organic movement really was talking about food production and not landscape plants. For example, you can't give phosphate to any plant from Australia. A lot of plants don't like organic matter. They want lean, sandy soils. Mediterranean plants, for example. If you gave them organic matter over a long period, you'd just be wasting your time.

I like to tell my students, "Plants talk to you. Now, listen carefully. They talk to you through chemicals, through color changes, through wilting. So you've got to look at your plants constantly because they are telling you whether they're healthy are not. The whole secret to great farming and horticulture is reading your plants. That's what it's about." Observations. You've got to be an observer. I would say to my students what Joe Connell told me, "I want you to sit in front of that flower for ten minutes and I don't want you to move a muscle. Just stare at the flower." At first nothing happens, but after about five minutes these insects start coming to the flower. They're drinking water and copulating and eating and looking for pollen and nectar. And the students come away feeling like they're part of it, not on top of it. I say, "Now, that's the feeling you have to have to grow. You have to have that feeling that you're part of that. If you can't do that, and if you see problems as problems instead of challenges, then you don't belong in horticulture. This is not for you."

[Teaching horticulture at a community college] was a wonderful challenge. It took me years to figure it out. I was an idiot the first few years. But you figure it out after a while, and you find those little life experiences and then they go, ah! You can't pay me enough for that. The look on their faces. It's just amazing to see that light bulb go on.

The hands-on thing—without that, you're dead in the water. In the organic gardening class I taught back in the seventies, we had this compost pile towards the beginning of class, and people brought garbage, and I showed them how to stack it: carbon, nitrogen, air, water, and microbes. Then we'd get cheese and wine to show them microbial cultures. This is a microbial culture that's just like that. It's just soil cheese and soil wine. At the end of the class we would turn the compost pile and reveal it, and there was this black, moldy stuff in there. [One woman] had been on my case the whole time. Her husband worked for Dow Chemical. I have a long fuse, but when it's lit, it's lit. So I kind of got in her face. I said, "Listen to me. Come here. I want you to smell this." She says, "God, this smells just like dirt." And then, her face. She got this red glow to her face and her eyes just lit up like this. And she said, "Oh!" At that moment, it was like, gotcha. That's teaching. That's what it's all about. As a teacher, if you get ten people in your life that just go, ooop! That's all you need. That's why I love teaching, because there's no substitute for that. Doctors probably have the same thing, and lawyers, in their own way, helping people.

A teacher is basically a student one step ahead, so I never left college, ever. I had never had a botany class. I'd never had a horticulture class when I started teaching horticulture. But I had a lot of biology, so I figured, well, what the heck. The first three years I was up every night writing lectures, making mistakes, low enrollment. Then after about three years you get these really good notes and you start getting little stories and little reference points. And you've still got your notes. Then after that, the notes start to not be important anymore. You can not look at your notes for a half an hour. By then the stories start taking on color and experiences that you've had and other people have had. Then you start involving other people. And then at the end you're performing. The notes are in the trash can and you are on, and you've got a tape in your mind that's just spitting information, and every story is a jaw-dropper, and people are laughing half the time. You're there. Thirty years of this and you get pretty good at it. There is nothing like that. Nothing like that, when people go, "Whoa! I got it." And you go, "Yes, I know. Of course you got it." "Whoa!" Welcome to the club. Because you know, after that nothing will ever be the same for them. That to me is the essential part of being a teacher. You change the way people see things. If you can do that, that's all there is to it.

Let me give you an[other] example of hands-on teaching. When I was at the El Mirasol project in Santa Barbara, which was an environmental garden in the middle of the city, we were going to put a farm there.[3] Also alternative energy. We had a methane digester and we were working with a guy named John Fry, who had one in South Africa. We built one using an inner tube. So you take a tractor inner tube. Think

[3] See "El Mirasol: Life on a Polyculture Urban Farm," Marjorie Popper, *Santa Barbara Independent*, January 3, 2008
http://www.independent.com/news/2008/jan/03/el-mirasol/

of your gut as a digester. You start at one end and out comes gas and food, which is basically sludge and methane. Okay? But wrap it around itself. Now you got a tube, right? We're raising chickens, right? Bring kids in there. We take the chicken manure. You put it in the digester and it goes around, right, and it digests. It just takes a couple of weeks, but you've got manure in there anyway. They put the chicken manure in one end and they go to the other end, and it's these fancy tubes and gadgets, and all sorts of stuff, and bricks holding it down to make pressure and everything. And out comes the methane at the other end. And that's attached to a stove and on the stove is a cup of water and in the cup of water is an egg. Then they cook the egg. And they take the egg-shells and they crumble them all up and they take them back and they feed them to the chicken. So in one brief moment they see the process.

See. That's what you've got to do in teaching horticulture. You've got to set the stage so that the students see a process. My whole adult life was challenged by trying to figure out ways of teaching process. The garden happened to be a tool. But it just was a tool for me. So I learned how to use the tool. But that wasn't the goal. If they learned plants, fine. But that wasn't what I was about. I was about process.

What's happening now happened in the early seventies. History is not a cycle. It's a spiral. It comes back on itself on a different level. And people my age now are walking around in a daze. Déjà vu! What do you mean? Like this couple that just came out with a book [on eating locally] and now they're famous. They are basically saying, "Well, our family decided that for one year we are not going to eat anything past one hundred miles from our house." They wrote this book on it, and they showed how to garden, and it was a nice little book. But there were scores of books like that written in the seventies.

Of course it's all just sour grapes. I understand that. But when you get to this age, you realize that history really is a spiral and we're just coming up. The difference now is it's global. That is the existential change. That is the—I hate that word—but the paradigm shift is this global consciousness that's beginning to emerge about, well, we've got to do something. And anyone that isn't doing something has to be ignored. Now, the thing that upsets me is that most fundamental religions are based on Armageddon and the future is tenuous. So how do you run a world with a tenuous future where people believe in Armageddon? This is the battle, as far as I'm concerned, of the future: putting people in power that actually see the politic of the future and not just the present.

Organic, organic, organic. I walk into a supermarket for the first time and I always ask for the produce manager and I say, "Where is your organic produce?" I can't see it and it's probably not there. But I want them to know that I want to know where it is. I've been doing that for thirty years and I'll be doing it the rest of my life.

I was in a supermarket in Aptos and I heard this scream. So I rush over to the produce section and this woman who was rifling through the lettuce saw a caterpillar in it. It was the organic produce and the store had just brought it in. I said to her, "Is everything okay?" "Well, that's a caterpillar." "Yes. That's okay. It's just a caterpillar. It means there're no pesticides on it. This is a good thing." "Oh, no." I said, "Ma'am," and then I launched into my spiel which gets me in trouble. I should have just walked away. No. I said, "Ma'am, would you rather have a tablespoon of DDT or would you rather eat that caterpillar? What would you rather do?"

Steve Gliessman

Photo: Jennifer McNulty

An internationally recognized leader in the field of agroecology, Stephen (Steve) Gliessman is the Ruth and Alfred Heller Professor Emeritus of Agroecology in the Environmental Studies Department at UC Santa Cruz. He earned his doctorate in plant ecology at the University of California, Santa Barbara and was the founding director of the UC Santa Cruz Agroecology Program (now the Center for Agroecology and Sustainable Food Systems). His teaching focuses on agroecology, sustainable agriculture, organic gardening, ethnobotany, California natural history, botany, and ecology. He is the author of the groundbreaking textbook Agroecology: The Ecology of Sustainable Food Systems *(Second Edition, CRC 2007), and numerous other books and articles. Gliessman founded and directs the Program in Community and Agroecology, an experiential living/learning program at UC Santa Cruz. He heads UC Santa Cruz's Agroecology Research Group, an interdisciplinary body of faculty, graduate, and undergraduate students, research associates, postdoctoral researchers, and international visitors. In 2001, Gliessman and his wife Robbie Jaffe started the Community Agroecology Network (CAN). CAN's goal is to help a network of agricultural communities in Mexico and Central America develop self-sufficiency, sustainable food systems, and local livelihoods. CAN and PICA also cosponsor the International Agroecology Short Course, which Gliessman has taught since 1999 in venues as diverse as Costa Rica, Mexico, and Vermont.*

I applied for the Organization of Tropical Studies course in Costa Rica for [the summer of 1969]. The people in the course were by and large fairly progressive environmentalists of that late-sixties time, wondering about what was going on with nature, why are forests being cut down? There was a lot of discussion. A lot of it was just, gosh, the tropics are an incredible place. There is all this diversity. Why is it that way? Why is it so different from temperate regions?

I took a quarter of Spanish when I found out I'd gotten accepted, so when I went down there, I almost immediately started hanging out with the Costa Rican driver, the cook, the field assistants. By the end of the time I was down there, I could pretty much carry on a conversation in Spanish, enough to get by.

I really connected to the tropics, learned so much, just drawn to it, but also connected to the people who lived there in the tropics. Not just the forest, but the people next to the forest who were trying to make a living.

About a year later, [my professors at UC Santa Barbara], Bob Haller and C.H. Muller, put some funding together to send me back down. I'd picked up an old used Land Rover. My [then] wife and I drove from California down to Costa Rica. We ended up spending three or four weeks doing a project on bracken. Found a place in the central part of the country where a mixed tropical evergreen-oak forest had been cut down—a lot of it turned into charcoal, sometimes planted in crops, a little bit of coffee—but a lot of it just turned over to pasture, degraded. Bracken was moving in and taking over and not letting the forest come back. I worked on the mechanisms of that, did some field sampling. We stayed in a little two-room hotel upstairs in Santa Maria de Dota, a little tiny town square. I'd go down the stairs and through the kitchen and out back where there was one faucet with some water, and I'd wash things out and then go back upstairs. People would wonder, "What the heck is he doing?" Doing little bioassays, doing a research project.

Driving out of the community you look down into a little valley and a little town down at the bottom with a town square. Everything from forest to pasture is around. I'd been struck while I was there about how weird it was that here I was studying ecology, about what makes nature work, and all I was seeing was the destruction of forests. I was seeing farmers do things that were causing them to have to abandon the land and move on and cut more forest down. It didn't make any sense. It seemed like ecology should be able to help. I mean, we work with soil, plants, animals. It's about what makes nature work, what allows systems to recover from disturbance, and here were these farmers doing every-thing wrong, at least it seemed like that's what was going on. I didn't know much at that time but that's just what struck me. I would think, "Gosh, there's got to be something I can do."

In 1972, I was invited to come back down to Costa Rica and teach in the OTS [Organization of Tropical Studies] course I had taken in 1969, even though I was still not finished with my graduate work; I was probably six months away. I accepted it, but by this time having thought a lot about what does ecology mean for solving problems. Earth Day had happened in 1970. At first I was sort of turned off by Earth Day because it just seemed to be recycling and tree hugging and no ecology. Us ecologists were sort of resistant and hesitant, and in fact downright negative to it. We said, "Well, there's no ecology in this." We weren't very supportive, at least those who were trying to make a name as scientific ecologists.

When I went back down in February of 1972 to teach the course, I was taken on as a botanist, a plant ecologist. I would go into the forest and do ecology there, and look at relationships and complexity and diversity and succession and mutualisms, and all the good stuff that happens in a tropical rainforest system. I wanted an experiment. I wanted to see if we could have some interaction with farmers and how ecology and agriculture might interact. I think one of the first places we went to, there were some little patches of forest that had been cut down for crops. We would break up into four groups of five and go out with a different faculty member to do a field problem, testing some ecology concept or principle. I decided to go to this farmer's field and try and understand how the soil was being managed, or why the farming was being done in the way it was. The first thing we did was we all picked up machetes and helped him clear, and got blisters, and one of us cut ourselves. But we heard the story of how you prepare the land, how you burn, how it fertilizes the soil, how the crop responds, how you get a good yield—this whole story of ecological management. That's basically what it was. Because it wasn't fertilized, it wasn't sprayed. It was using the machete, and fire, and seed, and human knowledge. I was struck by that, because my impression of agriculture had been you drove a tractor, and you sprayed, and you did all that kind of stuff, or you weren't a farmer. But here it was, this traditional farmer raising a crop with nothing except knowing how to do it.

So we were there for two months for this course, and I'm cogitating this, thinking it over and wondering about it. I had a couple of other opportunities in other parts of the country under different kinds of conditions to ask the same kinds of questions of farmers. We were towards the end of the course, and I'm really thinking, gosh, I want to do this again in a coffee-growing area. Because in those days in southern Costa

Rica they just cleared out the forest underneath a little bit and put the coffee under the forest as it was originally evolved to be, an understory species. I knew I wanted to take our group and look at that from an ecological perspective, see what it was like. It was a Sunday when it was my turn to take a group out. I asked the driver and the cook, "Do you know of any farmers in the area who we could go visit for the morning and help out and learn from?" He said, "Well, it's Sunday. No farmers are working today." But then one says, "Ah, but wait a minute. I know a gringo. His family has a farm about nine miles away, and he's always working. Even though it's Sunday, he'll be working." I said, "Great. Let's see if we can go over."

So we did, and he [Darryl Cole] was working. He showed us what he was doing. It turns out he was the son of a couple who had moved down to southern Costa Rica in 1951 and homesteaded. They cleared land and planted coffee and put in some crops. And here it was, 1972, twenty-one years later, and they're still there, and they're farming. He'd just started a new project. He'd been doing a lot of reading, a very self-taught guy, never finished high school in the classroom, was just at high-school age when his family moved there. They were starting to build a terrace system. He'd read books about organic gardening and farming and soil conservation and decided that he was going to try to build this terrace system, because it was all sloping land. He had gotten the idea from some of his reading to plant along the edges of the terraces a grass that was nonaggressive. It wouldn't spread out into the nearby area. It was a grass to cut and feed dairy cattle.

So he was building this thing. He hadn't planted any crops yet, but he was telling us how he wanted to rotate crops and not just have one crop, and they were going to grow the vegetables up there. He was moving out of coffee into vegetables, because coffee was harder to make money on. There were ups and downs in the market even those days. They are up in the mountains, and down in the lowlands below us there, a couple of hours away by road, were United Fruit banana plantations. A lot of workers lived down there, and they couldn't grow [these vegetables], but they ate them. He had the idea of taking the vegetables down there to market them. We spent a morning walking around the farm. He still had some parts that were native forest that he hadn't touched, and even today those are still there. We heard about the difficulties of maintaining soil fertility with all that rain, 250 inches of rain a year, the problems with pests and weeds and all that, and how to make a living. We kept asking questions that showed our ecological understanding, but there wasn't much agronomic understanding.

Towards the end of our morning, we were gathered around with him, and I asked him, "Well, from your perspective of having lived there for twenty years and trying to farm for that period of time, what would you recommend that we as ecologists could do to help solve the problems you face?" He looked me square in the face and he said, "Come down and join me. Put what you're learning to use here."

I thought about it. And I went home, back to Santa Barbara, finished my degree, took a postdoc for two months in Bloomington, Indiana to be part of a large study where they had taken a piece of farmland that had been carved out of maple-hickory-oak forest in southern Indiana, and farmed for a while. What they were going to do was take it out of farming and study the successional recovery process in a whole-system study where everything would be looked at, soil, plants, bugs, biomass, the whole thing. I would be in charge of the weeds/allelopathy part of that and see how much of a factor that was in this recovery process.[1] Interestingly, it was trying to look at how does agricultural land recover from a period of use, of disturbance. It was all contingent on getting a big NSF grant that would fund this thing for many years. I would be one of the postdocs on this study, with the hope that maybe eventually it would lead into a faculty position.

Well, after about two months I was not very happy with the project, or just the situation. I remember spending a lot of time in seminars listening to people sit around and try to impress the rest of the people in the room with how much they knew, something that still bothers me about scientists, especially in university settings. A little personal bias, or issue. But I kept hearing this invitation to make what I was doing meaningful, rather than just argue and be kind of theoretical.

I decided, I'm going to take that job in Costa Rica. I left that position in Indiana and went south. And gosh, from the minute we arrived at the farm, I loved it. I just jumped in and started doing everything you do on a farm. But I also would do different treatments. I'd do a cover crop, or I'd lay down a mulch, and I'd put down quadrats and sample

---

[1] Allelopathy is the chemical inhibition of one species by another and is the focus of Gliessman's research on bracken ferns beginning in the 1970s. See, for example, Gliessman, S. R. (1976), "Allelopathy in a Broad Spectrum of Environments as Illustrated by Bracken," *Botanical Journal of the Linnean Society*, 73: 95–104. doi: 10.1111/j.1095-8339.1976.tb02015.x

the amount of weeds, and put out chicken manure and compost in one and not in another, and compare growth rates. Just started to gather some numbers that, yes, you could really change things using ecological understanding and organic principles combined.

It turns out that there were a couple of other people thinking about [these issues] in OTS, a couple of the more radical-thinking ecologists at that time. I was in touch with them through the course I'd taught and the course I'd taken. We'd talked about, could you do this kind of stuff or not? Some of them today still do that kind of work, John Vandermeer at Michigan being one of them. Others are scattered around still. There was one guy especially, Steve Risch, who was already questioning ecology for being so separate from people and culture, and social issues, and trying to just study nature and not be of value to people.

We were only nine miles away from one of the OTS field stations. I immediately made contact with them, for a couple of reasons. One was to sell our vegetables to the students when they come to the field station, but two, letting them know that our farm was there for them to do experiments on. Some of the first experiments we did [were to] plant broccoli seedlings in three different places: inside the forest, at the forest edge, and out in the field, and just see how they did. Which one was attacked more by pests? We had all these beginning ideas about how agriculture worked from an ecological perspective.

So there was some sort of academic link there. In 1974, after I'd been at the farm for a year or so, I'd heard about, through the little network of scientists that would go down there, that there was going to be the first annual meeting of the International Association for Ecology (INTECOL). They were forming a new society and a new journal. That year the first issue of the journal would come out, called *Agro-Ecosystems,* which today is *Agriculture, Ecosystems and Environment.* INTECOL held its first international congress in The Hague in Holland. I decided to fill out the paper presentation request. I wanted to talk about what I'd learned in two years of managing the farm, where I saw ecology working. And lo and behold, it got selected as a plenary. I had never gotten up and talked in front of anybody like that before ever! And here I am at this big congress getting up and talking about ecology and agriculture.

For me, that [time] was such an education. I remember going to The Hague and making my presentations and sort of being scoffed at by the conventional ecologists. But a good number of people at that time who were becoming socially conscious and wanted to see ecology make a difference came to me and said, "This is a good idea. This is what we want to do with this new journal. This is why we are doing this." [The conventional ecologists had said] "You can't do ecology in a farming system. It's too disturbed! There's no way you can do ecology there." There was just this small group on the side that was saying, "Now, wait a minute. If all the elements of an ecosystem are there—" So I went back, and for a couple of different reasons decided to leave the farm.

[Next I got] a job in Guadalajara, Mexico, at a big nursery business that was trying to gear up for export of ornamental plants to the U.S., but had a growing business to little nursery outlets in Guadalajara and through towns up the coast through Culiacán, Mazatlán, Los Mochis, Hermosillo, Guaymas—all up and down the Pacific Coast of Mexico. So I jumped into this as the general manager responsible for everything from programming our plantings, and propagating, and purchasing, and selling—setting up a whole business in Mexico. And for the next three years that's what I did, ran this business.

Towards the end of those three years I ended up giving a couple of little talks about ecology, and before I knew it, I [got a job at the Colegio Superior de Agricultura Tropical in Tabasco, Mexico]. Off I went into this school that had started a year, two years before. It was meant to train folks from the tropics about agriculture. Where most everybody had been trained before was in a big university of agriculture in central Mexico up in the highlands, very much affected by Green Revolution technology. That battle was going on right then, and people were coming from another place with imported technologies. And it wouldn't work, so they got this idea, "Well, let's do it within the country, and within the region, and try to adapt it more locally."

I started teaching ecology to agronomists and doing studies in the field. Some of the studies were very conventional, in the sense that, okay, here's a field on the experimental grounds of the college where there's been tractor cultivation for several years. Let's put mulch down here, and not here, and see what happens. We were in the middle of a big, internationally funded development project where they had cut the forest down, built little communities for the people and told them, "You're going to farm conventionally because this is going to be the breadbasket of Mexico that we're going to export to the rest of the world." They thought a school within that area could help solve the problems that they faced.

But around the college and on the road from the little town Cárdenas, Tabasco, out the twenty-one kilometers to the school and the experiment station, you'd see alongside the road these remnants of traditional agriculture: home gardens, corn. And if you looked close, you'd notice it wasn't just corn, but there were beans and squash associated with it. There was this whole other kind of agriculture. And between this guy that hired me down there, Ricardo Almeida, and another guy over in the plant pathology department, Roberto García, who was also very much aware of ecology, we stopped at some of these traditional farms and really began to realize that from an ecological perspective those made a lot more sense than the conventional approach, and the big conventional project that mostly was going on at the experiment station.

So we began to study the ecology of these traditional farms. This started in September of '76. And boy, it sure, real quick began to become clear that there was an inherent ecology to those systems. They'd been around for a long time. For us, it was a matter of starting to take them apart and see, but also to get to know the people who were part of it, because you couldn't do it without them. They were the ones who had the experience and the practice.

So gosh, all sorts of stuff we began to study. We even used that traditional knowledge as a foundation for designing some of our experiments. We did experiments on their fields, off the experiment station, using their varieties but taking them apart. Like corn, beans, and squash—setting up an experiment on a farmer's field with his varieties and his planting practices with the intercrop as he would do it as one of our experiments, but the monocrop as other treatments, the things planted alone—and look at how they responded and what did best. It was obvious that here was something that we began to call *agroecología*.

I remember driving back up from the lowlands with Darryl [Cole] after one of those long market days. We'd had a good market day and things had gone well. We talked on our way up the mountain about how important ecology and agriculture were, and that we were onto something here. This is important to really show how the ecology of agriculture could work. I called it the ecology of agriculture. Going down to Tabasco and seeing the traditional farming systems, especially the Mayan systems, and getting to know some of the students who had come from those communities—most of them had that in their roots. There were kids of farming families scattered around the lowlands of Mexico. They all had that kind of experience. But they were being taught that there was a different kind of agriculture, that that's what was valuable, not what they'd learned growing up from their parents.

We were learning from them and showing how ecology and what they knew could be blended. Both Ricardo Almeida and Roberto García, the two folks that I worked with there, they knew this, but they hadn't had a way to articulate it very well. And me coming in with a strong background in ecology was able to pull that together, and we started teaching *agroecología*.

At the same time, there was a guy up at the National Postgraduate School of Agriculture, next door, practically, to the Undergraduate School of Agriculture in Central Mexico at a place called Chapingo. Efraím Hernández Xolocotzi, an old ethnobotanist, agronomist who, I think, was also part Native, Pueblan Indian. He'd been battling right and left to try and stop the Green Revolution in Mexico, and had been doing it by showing how important traditional farming systems were. By the time I got there, he was becoming kind of cynical because no matter what he did nothing stopped. We teamed up. We did several things together, ran some seminars, did some symposia. We talked about the importance of traditional knowledge and the agroecological foundation of that. It was fun putting all that together and seeing it evolve.

So when in 1978, two years into my time in Tabasco, the next International Association for Ecology Symposium happened in Jerusalem, I went and talked about the importance of traditional knowledge for developing sustainable farming systems in the tropics and the ecological foundations of that knowledge. So it was fun, and a whole different sort of sense in 1978 than in '74. And for several years, gosh, we had a great time. Big projects and senior theses and all sorts of good stuff, building agroecology, *agroecología*, spreading it around, writing about it, starting to get some things out into the international literature about it.

The origin of the word "agroecology" probably goes back to a paper [by] a guy named Dan Janzen, [who is] well known in ecology, tropical ecology, [and] still lives in Costa Rica and had a lot to do with some of the forest preservation and restoration that's gone on, especially in Guanacaste, the northwest part of Costa Rica. He published a paper called "Tropical Agroecosystems." He was mostly talking about how we need to look at agricultural systems in the tropics as ecosystems.[2] That was a big step, to get people to begin to think about an agricul-

---

[2] Dan Janzen, "Tropical Agroecosystems," *Science* (1973), 182:1212-1219.

tural system as an ecosystem. That was a pretty important paper. I cite it even today as a landmark.

In the summer of 1978 we organized what I called an intensive short course in tropical agroecology. Down there, July was our summer month, not really summer (it was the wet season), but when we didn't have classes. So we brought together folks from all over Mexico, and a few folks from outside that I knew, going way back to the days with OTS down in Costa Rica. We organized this course that was mostly taken by Mexicans, but we advertised it and we got some Central American participants as well.

It was a four-week intensive field course. Every day we would go out in the morning and do fieldwork and come in in the after-noon and process all we'd done, and then in the evening have lec-tures and presentations. This went on for four weeks. God, we did crazy stuff! We really had a good time. But boy, people were worn out by the end of it. I remember things like going out at night and putting up mist nets to catch bats, and then working these bats out of the nets with gloves on, making sure they don't bite you, because there're vampire bats and stuff down there too, and all the bats car-ried rabies. We wanted to see who was out doing what, and if they had pollen on their noses from different plants. Imagine being out at night in the tropics with gloves and a headlamp, and you've got to hang on to this bat. You can't let go of it, because it will swing around and bite you. And mosquitoes are attracted to the light, and they're covering your face, and they're biting all over you, and you can't even bat them away. No pun intended. Yes, we did some fun stuff look-ing at the ecology of farming systems, really trying to do agroecology.

In that first course I taught in 1978, there was this person who had just begun teaching. I think she was trying to finish her Ph.D. but had started doing some teaching at a university in Mexico City called the Iberoamericana University, a Jesuit university. She was in the depart-ment of anthropology, but working under a faculty member, Dr. Ángel Palerm, who had been doing stuff similar to what Efraím Hernández Xolocotzi had been doing, but as an anthropologist. He'd studied some of the prehistoric remnants and information about what was agriculture like in these regions before the Spaniards arrived, and what did it mean in terms of the evolution of agriculture in the area today. Fascinating stuff. She had worked with him, did her Ph.D. under him, doing studies in communities in several different parts of Mexico to try and get kind of a historical, anthropological look at what kind of farming systems had been there. When she heard about the short course in agroecology, she decided to come to it and participate, to learn the ecological side of some of the stuff she'd been doing. She'd been influenced, I guess, by some of the folks in anthropology who do what we call cultural ecology and wanted some more of the ecology. She hadn't been trained in it. And that's why she came, to see the interface between people and nature. She'd thought a lot about that.

Very soon into the course I realized that she wasn't just a normal student, that she was a faculty member, and that she was carrying a pretty important message about why she wanted to be there, and how to work ecology and culture together. I said, "Hey, I want you to present more of what you do and why it's important." And out of that we developed a relationship that continues today. Alba González Jacome. A phenomenal lady. We've had a lot of good times together and worked together on lots of different projects. I know that she probably, more than any other person, opened my eyes to the importance, number one, but the ways, number two, to link this nature-culture interface.

You can't just do agroecology on crops. There are people involved, too. I remember, in some of the early days of agroecology, trying to tell people that agroecosystems are more complex than natural ecosystems. And people would just laugh and say, "What do you mean? They're simplified ecosystems. There's nothing complex about them at all." I would try to respond by saying, "Well, wait a minute. You got the human factor that complicates things to no end. And how do you work with both at the same time?" Well, that was at a time, of course, when the Extension model was so predominant, where you develop technologies at the research stations and at the ag centers and you transfer it to the farmers. And if the farmers don't adapt it, it's because they're dumb, or backwards or— That was pretty interesting, when here we were trying to say that all systems are agroecosystems, but a lot of them have more or less culture as part of them. Look at these systems that have centuries of history under incredible conditions: pre-high technology, pre-fertilizers, pre-sprays, pre-machines—everything. The evidence was just starting to surface at that time too, that these weren't just scattered little small tribes of people. They were civilizations, highly developed, with very intricate social relationships, economic relationships, and farming systems that had evolved and developed in lowland tropical regions. It told me that that piece, the cultural piece, is essential.

A lot of people probably even today still have problems with the simple word "agroecology," because they can only think ecology, and they can't think the agro, bigger-picture food-systems part. When my first textbook came out it was *Agroecology: Ecological Processes in Sustainable Agriculture*. The second edition is *Agroecology: The Ecology of Sustainable Food Systems*, published in 2007. That's an intentional change on my part to try to explain that agroecology is more than just production systems. It's an ecologically based approach to understanding sustainability of entire food systems. It's as much a social phenomenon as it is an ecological phenomenon, but it's grounded in ecosystem thinking and interconnectedness, relationships, change over time—all of the concepts that are grounded in ecosystem thinking but in a human context.

Sean L. Swezey

Photo: Bev Ransom-UC-SAREP

As an entomologist and integrated pest management specialist, Sean L. Swezey has been a pioneer in organic farming research. With UC Santa Cruz agroecology professor Steve Gliessman, Swezey helped establish one of the first research-based organic farm advising services based at a university. He wrote and edited a University of California organic apple production manual—the first such manual for any organic commodity. He has also held a variety of influential posts, including director of the Davis-based University of California Sustainable Agriculture Research and Education Program (UC SAREP), research and teaching appointments at UC Berkeley, Cornell University, and UC Santa Cruz. He also worked as a consulting entomologist in Central and South America with the Organization of American States and with the Food and Agriculture Organization, and has advised a series of California secretaries of agriculture on the implementation and enforcement of the California Organic Foods Act of 1990, now enforced under the USDA National Organic Program since 2002.

Berkeley in the early seventies was a very stimulating intellectual *world*. Oh, what a challenging world! I studied with a group of professors in the College of Agriculture and you could still take an undergraduate degree in the college. Agricultural science was phasing out. They were slowly getting rid of Cooperative Extension professors, and majors directly related to agriculture, and were moving them either to Davis or Riverside. But I was only vaguely aware of that struggle at the time. The College of Agriculture was still there and fortunately I studied with and worked for the biological control/

IPM [integrated pest management] group. I took classes from Robert van den Bosch, Carl Huffaker, Don Dahlsten, Lou Falcon, and Ken Hagen, and many others who influenced me—all very well-known mid-century scientists in biocontrol.

I majored in natural resources as an undergraduate and then I received my Ph.D. in entomology in 1982. At that time, UC Berkeley was *the* ecological school of biological control. I still pinch myself that I had that opportunity. Because it died. After I left school, the Division of Biological Control was slowly phased out until the facility was closed in the 1990s. They completely did away with it. But it had been a formative time for me: seeing that science was a social process, and I studied with a group of scientists who had a strong research-based position, which was very much antiunilateral use of pesticides and probiological control. All of my teachers were very important to me and they still are to this day. I feel we're carrying on that work.

At that time, "organic," and "sustainability"—these were not words in anyone's mainstream vocabulary. But biological control, alternatives to pesticides, and van den Bosch's very radical position on integrated pest management were current. For the past twenty years at Santa Cruz, I have taught many of the same principles I learned in Berkeley in my pest management course. The Berkeley school of biological control is still very influential in my courses and research.

My interest was always in the array of natural forces that can be used for pest control—how do you manipulate or manage them? It seems rather quaint, twenty-five, thirty years later, but that's what you did at the Division of Biological Control—I studied beneficial insects. The laboratory was in Albany [California]. The facility was called the Gill Tract. It was a big open space there. We had a lab; we had a quarantine. My major professor studied naturally-occurring biological control agents at the time. I published a three-year study with him on natural enemies of the Western pine beetle. I worked in a national forest in Northern California. I finished my fieldwork for my dissertation in 1980. So I spent nine years in Berkeley.

[After finishing my Ph.D.] I made a major life transition. I wanted to continue to teach and investigate this area of community ecology and pest management. When I was twenty-seven, a UC professor of mine, Lou Falcon, said, "Look, we have this Organization of American States IPM program that's starting up in Nicaragua. You should go down there and visit these people." Falcon had been involved in a United Nations Development Program project there in the seventies, and they were trying to revive it. There was tremendous enthusiasm for IPM after the revolution. I went down in 1980 to see what was going on there. The country was in ruins after the war, but at the National Autonomous University [UNAN] in León [Nicaragua] there was a group of IPM movement people. They wanted to rebuild agriculture in Nicaragua with IPM as a central idea.

I said, "I know next to nothing about your crops. I don't know anything about the country." So I came home and did eight months of intensive Spanish, an apprenticeship with Lou Falcon to learn the export cotton system, the IPM strategies. And then my wife [Christine Sippi] and I went back down to Central America. I taught and did research in Nicaragua for seven years. It was a master's degree program. The students had to come to class at night because they all worked in agriculture during the day. It was not a traditional program but an accelerated graduate degree course. I was an employee of the OAS for those years.

The most pesticide-polluted place I had ever seen in my life was Nicaragua at that time. It was a disaster. Going up against the most abusive, pesticide-dominated agriculture was quite a shock. But we threw ourselves into the IPM work. That experience taught me to never take no for an answer. Never give up on new research results if field results warrant it. The IPM paradigm and the organic farming paradigms have been validated. I became absolutely convinced that the community-ecology, applied-ecology paradigm of pest management is the correct approach. We must continually train people to do the implementation. Scientific paradigms are socially constructed. Paradigms conflict with other constructs and incentives out there. But IPM and organic agriculture are undeniably successful, efficient, work and can be improved to successfully work in the future. The peer-reviewed research record I tried to put together reflected the fact that conventional pest control can reform pest control in a dramatic fashion.

I believe there has been an historical struggle between the more bio-intensive IPM approach we teach here [at UC Santa Cruz] versus the conventional land grant universities' "wise use of chemicals and biotechnology" course. We really do differentiate ourselves here. We emphasize bio-intensive, biological control sources of mortality and pest suppression, natural control models—the kind of ideas that really work with the organic growers. We're very much a product of California's IPM history and the emergence of the organic industry here, and the fact that we're so commodity-specialized here that we've been able to go with biological sources of mortality and restrict chemical sources of control to the minimum, or none, or compliant with the national organic rule. That has emerged here far more successfully than in other programs. We have developed that over twenty, twenty-five years in an independent fashion. The Department of Environmental Studies has given us the freedom to do that.

I would say that in a lot of pest control many conflicts of interest emerge, or are structurally wired in. The fact that we don't come from that conflicted background has certainly been important to me. We've always agreed here that we don't prioritize proprietary research, whereas, my experience elsewhere in pest control research was that almost everyone is doing proprietary research one way or another. In entomology, that will influence you because there's more money in getting hooked into private or industry funding than there is to stay an independent and go to the nonproprietary funders and organic farmers. But we've stuck with that model and I think we've been very successful. We've raised millions of dollars of competitive grant funds over the years here for these topics and generated many practical results for the organic farming community from the research.

I remember meeting Steve Gliessman in about 1980 or 1981 at [UC] Berkeley. Steve must have been starting his teaching and research career here. I was aware of the fact that Steve had very interesting, novel ways of looking at things. I remember at the time, also, Miguel Altieri had just been hired at Berkeley. I was the student member of his hiring committee; I voted to hire him. Miguel's been at UC Berkeley for thirty years, I think. And meeting Steve at around the same time, it was interesting. [I was] thinking, who are these agroecologists? They're more general community ecologists, yet they're making a lot of statements about the ecology of industrial agriculture, for a change.

[Eventually my wife and I] moved to the city of Corralitos in Santa Cruz County. I think one day I said, "I'll look Steve [Gliessman] up, because Steve and I have similar interests," and Steve mentioned, "We need an entomologist for a study we're doing." That meeting engendered a very long-term relationship with Steve and Jim Cochran [owner of Swanton Berry Farm].[1]

Everyone said, "Absolutely impossible. You can't produce strawberries organically." I said, "Okay, I'll come do the pest management part of that study." That was the beginning of a very interesting research relationship. Steve was central in saying, "If we're really going to do peer-reviewed research that we can publish about the performance of organic agriculture, here's a model. What you do is side-by-side comparing." We worked that "conversion studies" model for ten to fifteen years after that.

In the late eighties, early nineties, we compared organic and conventional production, side-by-side for three years [at Jim Cochran's Swanton Berry Farm]. The organic techniques were in a developmental stage, but sufficiently interesting for me to go on and propose a similar study for organic apples. Steve and I had worked with Jim Cochran for a while, and he suggested to me that the next step was to create a formal organic research and outreach program in the local farming community. At that time, the organic industry was small and very "seat of the pants." We felt that no one was taking the organic methods from the [UC Santa Cruz] Farm and going down into the Pajaro Valley farming community to get research-based results. This is how I met the Riders, another legendary well-known organic grower family here. They were a multi-generational apple growing family, community leaders, and excellent horticulturalists. It couldn't have been a better match. In 1989, for unsorted organic tonnage for fresh market or juice the prices were really high and Jim [Rider] became interested. We proposed, funded, and published a long-term study of organic apple production on his farm.

We initially kept working on introduced codling moth pheromones and mating disruption as an experiment on Jim's farm in Watsonville. This was one little bit of proprietary research I did. It was so important to get codling moth pheromones registered as a pesticide in California. And they have, to this day, been allowed by

---

[1] See the oral history with Jim Cochran in this book.

the National Organic Program rule. There are very few synthetics allowed in organic pest management but the pheromones are compliant synthetics. We just wouldn't have good organic fruit without the mating disruption system. So it was a really good road to go down to make sure pheromones would be allowed. We now have apple orchards that are over twenty years organic under the mating disruption system.

The two Jims [Cochran and Rider]—these were the two people who were our anchor organic farmers in the community for research. They were open to setting up my research plots. I didn't find either one of them very skeptical about the eventual ability to succeed with organic production, which was very progressive for the time. Jim Cochran was growing organic strawberries and very few other people were doing that at the time. And then there was Jim Rider and a few farmers in Browns Valley [in the Santa Cruz Mountains] growing organic apples. There was nobody else. I eventually spent ten years in Rider's orchards on research projects. We held yearly extension meetings—the "Moth Madness" meetings. Sam Earnshaw of CAFF kept those going. We publicized our organic apple research results to the farmers each year. When you meet farmers like Jim, you admire their production records over the last twenty years. They're successful entrepreneurs. We were *very* lucky to work with them. In both cases, apples and strawberries, we published peer-reviewed results.

Apples and strawberries have to be [among the most] cosmetically perfect organic commodities we produce in the county, emblems of our organic fruit production. Now both are produced on thousands of acres here. I don't have the exact numbers, but, for example, organic strawberries grew from maybe possibly ten acres in 1988, to two thousand acres now in the state. There are over a thousand acres of organic strawberries close by here. And organic apples went from fifty acres, at one time, to four or five thousand statewide. Now, unfortunately, the apple industry is declining, so you can't say that any more. But we still have thousands of acres of each commodity in organic production.

In the case of organic strawberries here, the first research breakthrough was with beneficial predatory mites. We started that research with Jim Cochran and he became very skilled with them. Since 2000, our research has emphasized trap crops. Our newer recent research is on inter-planting alfalfa with strawberries. And in the case of apples, the breakthrough was pheromone-based mating disruption, without a doubt. However, *none* of those technologies was acceptable at the time of our first local research efforts. Now they are cornerstone organic technologies. So to make a long story short, the organic movement was very formative in the eighties and is now in full production with once "experimental" methods that are now accepted practices. But these methods were all things we worked on first as experiments. They were important new technologies that were not in practice but are now commonplace. It's a lesson: you can't imagine what organic technologies will be successful until you begin the pair-wise comparison evaluation of their efficacy.

We had some difficulties [getting our work published] in the beginning. I advocated to Steve that we publish in *California Agriculture*, which is *the* mainstream magazine for the Division of Agriculture and Natural Resources in the UC system. I believe it was the first time that anything serious with the "O-word" appeared in that publication. The challenge was, "[Here is] an extremely chemically-dependent crop. A UC [chemical] production system bundle has been created for it. And we're saying, Here is an organic production approach for it." I think UC reviewers looked at it and said, "What does this mean about our technical bundle if these people are saying, 'Well, this organic technique could be successful, too'?"

Slowly, during the nineties, we had more success with publishing in *California Agriculture*. I sent [an article] on organic apples; I sent one on organic cotton. In the late nineties or early in this century, we crossed a threshold where the organic research legitimacy struggle was won and now the task is creating a strong body of scientific literature to substantiate the claims of organic farming.

In the early 1990s, I met marketing people from Patagonia, an outdoor clothing company. They led me to believe that the next big organic commodity was going to be cotton fiber. And that led to eight years [of field research] with a group of organic cotton farmers near Chowchilla. Pesticide-dependent, a very problematic crop—can you produce cotton organically? The farmer that I met who was really producing organic cotton and really believed in it was Claude Sheppard, and Linda, his wife. We put together a study group. And for about six or seven years, they sold organic cotton. Patagonia bought most of it, or Esprit, and some other apparel companies.

Claude was certified to produce organic cotton. My biggest question as a research scientist was: Why in the northern San Joaquin Valley can you produce cotton without insecticides? And we found out. It was a fascinating story. It was because of the mixture with other surrounding crops but alfalfa hay was probably the big player.

We found out that as long as you had reasonable mite and aphid mite control, the only key pest was the lygus bugs, and they were in this special relationship with the surrounding alfalfa fields. We were only vaguely familiar with this kind of ecological situation. But Claude Sheppard's grandfather and father had observed the phenomenon. In fact, most of the conventional growers we compared Claude with also were very low insecticide users. They all agreed: "You can really grow cotton without insecticides around here."

So we formed the adjunct scientific team of a growers' group called BASIC, Biological Agricultural Systems in Cotton. BASIC was started by [Sustainable Cotton Project founding director] Will Allen. The EPA awarded us funds and the California Energy Commission awarded us research money. We published. Organic cotton yields were about 10 percent, somewhat lower than conventional, so organic cotton needed a premium price. Conventional cotton seed is now all becoming genetically modified Roundup Ready anyway, so we were slowly surrounded by GMO cotton plantings. That was an interesting thing that happened in the 2000s; everybody started planting these Roundup Ready cotton varieties, and we were the only ones sitting out there with this organic idea. So basically we did some of this research with transgene varieties all around us.

We worked with the Sheppards from 1992 until 2001. We published some early research in the mid-nineties in *California Agriculture*, but then we published a paper on: What's the long-term viability of this crop? We looked at long-term yields; we looked at long-term insect control. We looked at costs. And it turns out organic cotton is competitive if you can get a price premium and has fascinating environmental dividends. Organic cotton really took us by surprise. People buy the crop because it isn't associated with cotton pesticide residues. There was a marketplace for it.

Well, then organic cotton prices fell dramatically. You learn, when you start to expand organic production, that cotton is traded in the international marketplace. We have gone through these cycles of lower international prices. So it wasn't viable for the organic cotton growers any more. They couldn't really command a sufficient premium for their costs. Now there's no organic cotton in California, so there is no place [for us] to work. International producers like India and China are putting organic cotton on the world market at a price we absolutely are not competitive with. In that Chowchilla area, cotton is being replaced by almonds and dairy.

Organic cotton is here as a commodity now. But I go into Walmart, and I see the prices of the baby clothes. I can see all the commodities that are made with organic cotton, and that's not California cotton. It is imported organic cotton. There's a farmer in Texas that works in the High Plains that I think has it pretty good there, the Texas Organic Cotton Cooperative. So there're some organic cotton groups still out there. But the California organic production cotton production was dead by about 2002. We published our long-term research study and said, "Well, if it ever comes back, we'll work more on the production practices." I've had a few calls from farmers and marketers every once in a while: "Can we do this?" Not at our prices. I just don't think it's competitive.

Fortunately, that hasn't happened for strawberries and apples—although apples, because of the land values around here, it's a dying local industry. The Riders still independently pack and are profitable. But the cotton problem was an eye-opener. After ten years of working on it, we got world prices in the face.

Scientific research work on organic production is much more specialized now, the scientific work on organic. We're asking more precise questions on how pest control works in organic production, rather than these more general conversion studies questions. We've been more focused on peer-reviewed science with the participation of [organic] strawberry growers. I have a long-term project to work with strawberry growers who want to plant alfalfa-strawberry polycultures.

What we learned in the cotton system was that adult lygus bugs fly between fields. Lygus bugs damage cotton buds and also damage strawberry fruit, fundamentally displaying the same ecological role as a seed or bud "predator" in each system. We knew that the lygus bug was a major problem everywhere in Central Coast strawberries. It doesn't make a difference whether the field is organic or conventional. Lygus bug is a native insect. It flies in from the surrounding vegetation and around July it will damage the strawberry crop. Have you ever seen a summer strawberry that looks seedy—all twisted and funny? Have you ever wondered what does that? It's lygus bug feeding. It feeds on the face of the immature berry. Then the berry develops as a deformed fruit.

The consumer expects an organic strawberry pack that is full, well-shaped fruit. You can't put twisted, seedy fruit in the pack. You might have a little leeway in some markets, like CSAs [community supported agriculture] or something like that, for smaller fruit, but not in retail commerce. Strawberry growers are getting pretty big. We've got a national organic strawberry marketplace now.

We had observed that you could put a far more attractive host plant out in the fields and trap [lygus bugs]. This is called a trap crop. Claude and Linda Sheppard had explained this to us. We knew that lygus bugs spend time in the hay fields surrounding organic cotton, and it was only when you really disturbed the hay field by mowing that they would fly over to the cotton field, if they have a choice. In the scientific literature, it's been known for over fifty years. I had to go back and be reintroduced to the idea. But then we decided, okay, we're going to try to apply this principle in strawberries.

We started in the late nineties. Ten years we've been doing this work, and we finally came up with a system that really works. Most people think, mix flowers with your crops, that's good. But we really want to know which one's the best one and how you manage it. Without a doubt, alfalfa is the best trap crop for lygus bugs in the strawberry system. We have peer-reviewed work published on our experiments. I was just presenting results at the Entomological Society of America meeting, the national meeting in Reno. We're routinely putting up posters and giving talks about it at national research meetings.

So the eventual outcome of all of the work we did with [strawberry farmer] Larry Eddings at Pacific Gold Farms is: he, to trap lygus bugs, devotes about 2 percent of his fields to alfalfa. Forty or fifty strawberry rows are planted for each row of alfalfa; then you'll see a row of alfalfa. Larry learned how to manage the alfalfa. He has a tractor mounted with a vacuum in the front. A fan runs the air upward so that you can run it over the alfalfa and remove the lygus bugs. They're called bug vacs. The technology is kind of variable. We used an old technology, these vacuums that came in in the eighties, but that had been largely abandoned. We now have good efficient vacs that work well with this particular alfalfa-strawberry bundle planting. A farming company has adopted this technique and several other farmers are doing it. We're slowly getting information out to the farming community about this approach.

Another feature of my research has been recently published. I have a great research colleague who works for the California Department of Food and Agriculture in their biological control program, Dr. Charlie Pickett. He studied at Texas A&M. He and his colleagues documented that the lygus bug in Southern Europe has a sister species to ours in California, with very similar behavior in Spain and Italy, and it has a parasitoid that is very efficient, that doesn't exist here. Charlie and his colleagues imported and quarantined that parasitoid for introduction into California. He got the legal permit for importation. We released it here. The alfalfa trap crops now are anchoring large populations of the parasitoids. This successfully established parasatoid is spreading everywhere, attacking lygus bug nymphs in strawberries on the Central Coast. So the management of the alfalfa trap crops now has another purpose, which is retaining a very specific imported parasitoide there. We are documenting the spread of this parasitoid all around the original introduction areas near Watsonville and Salinas. So we hope that there'll actually be a bigger dividend for everybody, because this classical biocontrol parasitoids agent will reproduce on its own. It's out there. We're not going to release it again. It's doing its thing. It's adapted. We have just published data that show sixty or seventy percent of all the lygus bugs in an alfalfa trap crop are parasitized. The parasites are very successful.

Farmers who want to adopt this strawberry-alfalfa system are very interested. They have come from all over the world to visit us. We've had visitors from Italy, Spain, China, Canada, Sweden. There's a group from Ontario that's coming out this winter. I think the Canadians are the next group who are really going to adopt it. We host several meetings or tours a year, just to show scientists our trap cropping work. Believe it or not, you can fill a room with about fifty or sixty world scientists who are specialized in lygus bug research. The International Lygus Bug Conference Symposium meeting was here a year ago. We happened to have them here for a week in the United States at Asilomar. We took them out on the tour of trap cropped farms near Salinas.

Now Larry Eddings says, "Well, what's next? You've got the double whammy here. You've made lygus bug a very low priority for us. We want to move on." So we're evaluating the vacuum machines, how to run them more efficiently, how to use them less. We don't want to overuse the vacuum machines, if we can help it, because they're expensive. Tractors use diesel fuel. They disturb things. Air flow over a crop is like a pesticide, too. It's not selective. We want to be very precise with the use of all of this technology.

I also made a very interesting productive contact with another scientist, James Hagler, a well-known entomologist. He's at the USDA Arid Land Agricultural Research Center in Arizona. He invented an innovative way to mark the the trap crop so you can tell that the lygus bug was there. You can do it with egg and milk powder for marking in the field and lab antigens in the lab, assay the marks with an ELISA test, enzyme-linked, antigen-based system. Now we can actually go out and find out where lygus bugs have been in the system by spraying milk and egg powder for which we have antigens for a lab assay test. Field collected lygus bugs are brought into the laboratory, and if a prepared sample fluoresces, then that one confirms that the insect originated in a marked trap crop. So now we're doing flight-distance studies: how far do they fly from and between trap crops in an organic strawberry field?

So you see how the research has changed—from legitimacy struggle approaches on how to do organic, or side-by-side comparison studies, to now trying to work out mechanisms. I think the organic research literature is going to be more and more specific in the future, less and less: can you do it or not? Maybe even less and less: is yield exactly the same? Because as we get better and better, yields come up and prices fall. I'm seriously concerned about oversupply of organic markets. If we want the next leap forward, we need another ten to twenty percent of the consumers to start consuming organic product, rather than worry about trying to convince them you can do it. There's plenty of product out there now.

We believe that initial research questions should emanate from a discussion with the farmers, that, and you should formally have the farmer help the investigator put the experiment on the farm, that research within the constraints and conditions of the organic farm is much more realistic for results than isolated small plots on experiment stations that are highly artificial. And then finally, when the results are in, back to the farmer, explanation, discussion, downloading—we have all of these terms for it—and peer review with the farmers acknowledged as central players. Jim Rider was a coauthor with me on a paper once years ago. We haven't demanded that since. The farmers are too busy to do that kind of stuff. Certainly their in-kind contribution to our research is the central point feature of our existence.

Our model is distinct from most of mainstream agricultural research. It's not that novel any more. Certainly on-farm research has been an historical construct. We elevate the farmers, the organic farmers' field, and the organic farmers' interest, and the organic farmers' commitment. That's what's made us unique.

There are very few people who can make it all the way to peer review, because [this collaborative approach] makes a lot of the science much more difficult to analyze. Sometimes we have interesting results, but it's not quite publishable. Or, usually the farmers tell us they're five years ahead of us by the time our promising results are published. A very gratifying statement that the farmers routinely make to me is: "Sean, by the time you get that through peer review, we're five years ahead in our thinking about where we're going next." To me, that's the way it should be. Rather than the other way around, which is, "I'm not going to divulge it until it's reviewed," or, "You won't get anything until you have to read it in a scientific journal." We've never been that way. We've been quite the opposite.

It's a service. The farmers we work with know we will substantiate results immediately for them. If they have an absolute question, a statistical question, where the collaborating farmer who's working on a large funded project comes to us with an immediate question, we'll look at very mundane things very quickly.

We've neglected this whole issue of how much we need farmers. Rural people who have the production talents we study are like an endangered species. I sometimes go to the farm community and say, "Look, I don't want to put you under the microscope, but you're pretty important. You're a minority person in our economy, but your production activities are so important. We need to promote you, show you some respect." I think the general tendency in this society is—farmers are bumpkins. They're not modern. Most of the farmers that I encountered over the past twenty years were open, progressive.

They believed in generating research information. They believed in taking some risks. That kind of interaction with farmers for a scientist is good. There's value to understanding what's going on on our farms, because the rhetoric is empty without some good, tangible scientific results about alternatives. Without legitimizing the alternatives, all the rhetoric of sustainability is just empty. All these farmers in their own ways played an essential role.

Most of the farmers that I have worked with were conventional before they were organic, and then all of them transitioned to 100 percent organic during my research time with them. They taught me about persistence: persistence in innovative production practices, persistence in organic agriculture. Staying power. I always try to teach that to students, too. This isn't a casual thing. The farmers will support you if you show up and keep working and don't abandon a good idea prematurely. Keep working; keep working.

Orin Martin

Photo: Sarah Rabkin

Orin Martin manages the Alan Chadwick Garden at UC Santa Cruz. When Martin was a student at American University in Washington, DC. in the late 1960s, he "got politicized" by current events: some 100,000 citizens marched on the Pentagon to protest the Vietnam war; Martin Luther King, Jr. and Robert F. Kennedy were assassinated. In 1969, Martin followed some friends to Santa Cruz, where he heard about "this place called 'The Garden'"—the one being cultivated by Alan Chadwick and his protégés on the UC Santa Cruz campus. He attended lectures given by Alan Chadwick, began volunteering at the Farm and Garden, completed an apprenticeship in 1975, and by 1977 was hired to oversee the Farm and Garden. He is widely admired for his skills as a master orchardist, horticulturalist, and teacher.

Growing up, I thought gardening was an onerous chore and kind of sissy stuff, actually. I had absolutely no interest. I mean, I loved the outdoors. I was very physical and involved with athletics growing up and really loved the ocean and the woods of New England. But not gardening. [laughs]

I had some friends who were going to school at UC Santa Cruz. They said, "It's really nice here. Why don't you come out?" This was in 1969. I got a VW bus and drove out here, and there were two people living in [my friends'] house that worked at this place called "The Garden" on campus. They came in late and left before light. So after a while, I thought I would investigate where this garden place was. I wandered up there one morning and I was just bowled over, and fell in love with it, and felt, I have to do this.

It was the garden that we now call the Chadwick Garden. When [Alan] Chadwick was here, there was an apprentice program but there was nothing formal. He just started going at it and students flocked there. That was the style he was familiar with, and that was how he was educated as per horticulture, and also as per acting. In that apprentice model of learning, you work with those who have greater competency side-by-side, and then you are put through your paces. It really is about redundancy, and working in place, and rote repetition, memory, to achieve excellence within a narrow frame of things. It's not as comprehensive as the apprentice program curriculum is now. [Chadwick] taught a lot with metaphor and story, and was enamored with Greek mythology, and, as he called it, classical horticulture—Northern European market gardening/estate gardening type stuff, so I got a lot of that.

He was a very magnetic personality. When you were in the room with him, there was energy. It was a little scary, quite frankly. He would draw these kind of impressionable students in, kind of like they were going to be in the inner circle, and then, "You're forbidden to come back to the Garden ever again." I'm sure he had blood chemistry problems. He was up and down on an hourly, daily basis. To be quite frank, he was just a tortured individual and a terrible person in terms of the audience he was dealing with, extremely impressionable eighteen- to twenty-one year olds. He was often psychologically cruel. He was seriously misogynist. He would use terrible language and rail on women and felt they had no place in the Garden, with the rare exception of Beth Benjamin, who, along with Jim Nelson, started Camp Joy and is still around in the Boulder Creek area. He liked her. But it was very rare.

All the great things you hear about him are true, also. I just never could cotton to the way he treated people. The moment he met you he either thought you were great or you were terrible and there was nothing you could do either way. You could be one of the chosen ones and mess up, as it were, and you couldn't alter his opinion. He actually was always very polite to me and liked me, so I didn't have that specific issue with him myself. Yet, I didn't know why that was.

I was trying to decide—because Chadwick was [such] a mercurial personality—whether I wanted to apprentice with him. I would attend lectures and was kind of around the fringes of the Garden. By the time I decided I didn't want to apprentice with Chadwick, he had moved on to

Covelo.[1] Then I started volunteering at the Farm and Garden. Seventy-two is when, for all intents and purposes, the Farm started up. And I did that in 1972, '73, in a sporadic fashion, one to three days a week. And then Stephen Kaffka was kind of the heir apparent to Chadwick, and he was formalizing the apprenticeship through UC Extension. It was a year-long program then, and I applied. I was in it '74-'75.

Thirty-one years ago today [laughs], Steve Kaffka left, and Environmental Studies hired myself and "Big" Jim Nelson to run the Farm and Garden. They said they would pay us $600 a month and that we couldn't have an apprentice program. And there was, at that time, no machinery on the Farm. Did we want the job? We said, "Oh, yes. We'll take it."

In truth, we had an apprentice program in the fifteen months when the program was [officially] in abeyance. Basically, we started doing it, and the word got out and people started coming to the Farm saying, "Well, can we work here, and would you teach us stuff?" We said, yes. Then they'd go home at night. And after a while they said, "We come at 6 [a.m.] and we leave at 7 [p.m.]. What if we just kind of put our sleeping bags over there?" And we said, hmm. I guess in a sense you could say it might be looked at as irresponsible, but in those early years there were a number of instances where we adopted the thing of—what do they say?—"It's easier to say I'm sorry than may I?" And there wasn't as much scrutiny as there is now.

When I arrived as manager [the Garden] was on a minimal subsistence. Maybe a couple times a week people would go up there and water some of the perennials, and maybe there'd be a cover crop here or there, but it was just really on a low ebb from when Chadwick left in '72, to '77, when I started as a manager. The main garden, which is about a half-acre of raised beds, was gone. It was all just weeds. There were some fruit trees that Chadwick had put in that were tended and cared for in the kind of derelict years there.

We started the process of reconstructing the Garden over a period of years. Probably by '80-'81 it was flourishing again. Subsequently, in the eighties through the present time we've chosen to emphasize deciduous dwarf fruit trees. That's a driving thing that figures large in

---

[1] In 1972, Alan Chadwick was invited to Covelo, in Mendocino County, California to start an organic farm. For more information see the Online Archive of California archive of the Round Valley Garden Project Collection at http://www.oac.cdlib. org/findaid/ark:/13030/kt6q2nc89j/

the landscape and in the curriculum of the apprentice program, and some of our public service outreach, and writings and literature—to the point where we have quite a collection of apples. I actually don't know how many varieties we have, because we haven't done an updated inventory in the last two years, but I'm going to wager somewhere in excess of 120, maybe close to 150 varieties of apples. We've branched out and we have a sizeable collection of stone fruits: peaches, nectarines, plums, pluots and the like, and a pretty sizeable collection of citrus.

The way it's evolved is a half an acre plus of intensive raised beds with annual vegetable, flower, herb culture. A burgeoning rose collection. A number of years ago we got a grant from the Stanley Smith Foundation, which funds only ornamental horticulture projects, around organic growing of roses. They didn't care about the organics. All they cared about was the ornamentals. But what we signed on for, and I think achieved quite well, was to install a collection of roses. And one of the overriding things was to only use roses that backyard gardeners could readily get in retail operations, not all the great exotic old heirlooms. So that involved three classes of modern bush roses: floribundas, hybrid teas, and grandifloras.

And then to do public education about an organic approach to rose growing. They are probably the most disease-prone crop you can imagine on the planet, and Santa Cruz's climate really kicks that up a notch because of the cool coastal conditions. Fungal problems abound. I wrote a curriculum packet for the apprentice program that I also use with interns and student groups, around organic rose growing, and then we were supposed to write a pamphlet for the public. It kind of turned into, I joke, a pamphlet on steroids, or a "bookette."

Chadwick put in a few citrus trees when he first came, because having come from England, he was so excited about being able to grow citrus. And over the years, we've increased it too. We have twenty or thirty varieties of citrus. This is not the best climate for citrus, so the aim is basically extended varietal trials in terms of sweet oranges, mandarins—all different citrus classes—what will ripen in Santa Cruz? So we plant trees, and sometimes we keep them and sometimes we rip them out, based on the results. But we have a collection now and a database of recommendations.

We have a very eclectic group of apprentices. We have no educational or experiential prerequisites to get into the program. We're just looking for a good match: do they want what we have; do we have what they want? How are they going to use this when they go out from here to improve various sectors in what we're now calling "the agro-food system"? We draw—and now have a curriculum to reflect it—people who will be small-scale producers, farmers and market gardeners; people who will work in community gardens, community empowerment groups that want to have a gardening component, mostly urban, but some rural; and then people who are going to become educators. We have had people who parlay it into class credit associated with their teaching credential; people who are involved with environmental science camps, Life Lab projects, environmental educators. People who do development work overseas.

And then we have a fluctuating international pool of apprentices. For about ten or twelve years we had an association with the Margolis Foundation. Their goal was to fund study by students from Africa in agriculture, and they paid for two scholarships for the apprenticeship each year. One of our bywords for the apprentice program is, "We teach the teachers and we train the trainers." That's the multiplier effect from out here. And these students from Africa, it's really graphic. They go and they start small training centers that are magnets for people to come to. It's quantifiable how they increase food production and food security in the areas that they're in.

So you could say our mission statement is: we teach people to grow plants; the applications are many and varied. We certainly want them to have both practical and thinking skills, regarding—in the old days we used to say organic farming and gardening; now we're calling it sustainable production. Same thing, really. That's to be able to grow sustainably, whatever the scale—garden, farm—and whatever the crops. But more than that, to have been exposed to a mindset that allows them to analyze and think critically.

We want them to have the nuts and bolts of how-to, practical and high-order thinking skills regarding organic gardening and farming. And yet, it's more than that. They have an appreciation, respect, a sense of wonderment about the matrix of air, soil, and plants. We've always prided ourselves on being plant people, [with] a love of plants, wild and cultivated. Now, that's not something that will get you a job or anything, but that's part of the ethic. But the real key is that it's not just enough to do it. It's, how are you going to use it in a social context that, pretentious as it might sound, changes the world?

I don't think it's unique to us, but we have these phrases in our apprentice style of teaching where we say, "I do; we do; you do." We get their attention and we show them a technique and it's embellished by some concepts. Then we do it with them, and then we kind of push them out and see if they have the competency to do it on their own.

You say, "This is what we're going to do today." And there are some really strict and difficult-to-enforce time limits. You talk for no more than ten or fifteen minutes. You then make the students regurgitate, break up into small groups and spit back what you said, what are the salient points. And then you say, "I'm going to show you how to do this technique that's related to that theory," and you show it. And then they engage. That's just an example.

The program has changed radically since I first arrived. I would characterize it in the early years as more like the traditional old-style apprenticeship, where, in fact, we didn't have classes. Lyn Garling was one of the precipitators of classes, and I remember the incident. It was in '83, 84. We were out teaching in the field about growing winter squashes and pumpkins, and she's talking about the flower and the parts of the flower and pollination, and one of the apprentices looked at her and said, "Huh? What's an anther?" Lyn looked at me and said, "Orin, we've got to back some of this up with some classroom stuff." She had a master's in entomology and was the most intelligent person I ever met. She pretty much started to put it together herself. She just holed up and said, "Okay, we're going to have something on pathology next week so I'm going to go away for a week and [prepare]." She was bright enough to be able to absorb and spit it out. In fact, then she would connect with various academics and professionals in more technical, science-based stuff and bring them in. It was a huge watershed in the nature and the scope of the classes that we'd give to the apprentices.

There's so much jargon in horticulture and agriculture and the apprentices wouldn't know what we were talking about. It was like we were talking a foreign language, so we'd sit down and talk about the principles of soil science: structure, texture, soil profile, things like that. And then that led to, we should actually have some classes on soil science. So from the late seventies to the late-eighties, Big Jim Nelson—he was a man of prodigious skills, I might add, a great jazz piano player; he actually had a photographic memory and was quite a bright guy. He had a background in biology. He would teach any kind of technical class. We started to institute more science-based, technical type classes. We'd just do them as we thought it was appropriate, seasonally.

And then we started the schedule that we have now, which is that the apprentice program is in session from eight to six Monday through Friday. But on Wednesdays in the morning and in the afternoon we have two- or three-hour lecture-style classes breaking down the components of agriculture: entomology, pathology, soil science, etcetera. Then we have another series of classes that happen Thursday afternoon. We call them crop talks. They are largely taught by the staff and will be on garlic, on the tomato family, on crops.

About ten years ago we received a grant. It turned out to be a ridiculously small amount of money for what we actually committed to doing and did, which is we developed what we call a curriculum manual, which is not the entire apprentice program, but the outlines. It's gotten really good reviews. It's being used in institutions of higher learning. The presumption is that you already know how to garden or farm, but you want to teach it.

At any rate, the curriculum has become much broader and deeper, and well articulated, over the last twenty years, with associated readings as well. The apprentice program has gone more in the direction of the scholastic model. But again, we still put primacy on the practical. They spend hundreds of hours in the classroom but thousands of hours in the garden. So that's the balance. We are offering more classes and readings about social justice within the agri-food system now, in tandem with the horticultural and agricultural science, so it's changed quite a bit, although still the roots are in a practical training program—people who want to *do*.

There's a fellow at Harvard [named] Howard Gardner. He wrote *Intelligence in Seven Steps* and many books like that. He writes knowledgeably about the scholastic model and the apprentice model, and how they've mostly been on parallel paths, but occasionally very innovative schools over time have merged them. And that's what we've been trying to do at the Farm and Garden. We still put primacy on the practical. That's our focus. But in order to know how to dig a bed without wrecking the soil, you have to know some fundamentals of soil science.

And that has always been a bone of contention. We use theory to drive the practical, but we're practitioners, whereas the university is about scholarly research. I think [the reason for] a lot of the Farm and Garden's struggles over the years is just that we're misplaced. We can do what we do to the nth degree of excellence—I'm not saying we do, but we do a pretty good job—but it still isn't going to pass muster in the UC system. It's the wrong rubric. The apprentice program and the Farm and Garden would fit much better in the junior college or the California State University, CSU system. And that's insoluble, although we have a better relationship now.

Maybe I'm being kind of haughty here, but I think it's difficult within the context of the University of California to see that you can have fundamentals of science—concepts, theory—drive the thing, and still, it's farming and gardening. It's a craft. It's a skill-based thing. It's about a life lived in place. There's a great line from a Wendell Berry poem about standing in a field longer than a man's life. I mean, we're talking about soil, geological time to form it.

My kids have this great handbag, from the Whitney Museum in New York. It has all this wild drawing, and it says, "Art can't hurt you." And to me that really sums it up. You can talk cation exchange, base saturation, texture, structure of soil. And you need to. And you can also talk about soil in a poetic fashion.

I grew up in the sixties on the East Coast, and I studied a lot of Greek and Roman culture. [In the] Greek model of a citizen farmer, [both] brains and brawn were of value. You could do some things that were physical and you could have a rich intellectual life. Quite frankly, that's one of the things that's appealed to me about the context of growing things here in the university, rather than just being a farmer somewhere.

Chadwick was—let's face it, he was kind of a raver. He set people off. He and [Professor of Biology] Kenneth Thimann would have these debates. Thimann would be saying, "xylem flow," and Chadwick would be talking about "an exuberance of plant juices." If you took a step back and you listened to Chadwick talking about the hydraulics within the plant in terms of the water and nutrient movement from the soil matrix into the plant, out of the plant, and you listened to Thimann—they can't hear each other, but they're saying the exact same thing, and they're both right. But they're just so hung up on their different cosmologies, paradigms, whatever you want to call it. It was too bad. Of course, personality always factors into that, and they were two strong personalities. The camps in the early years, they kind of became entrenched. It was classic miscommunication.

I think one of the things that has made us effective educators is that we are passionate about what we do. Apparently, that is visible and palpable to students. They seem to respond to that. The curriculum is the Farm and Garden. It's the most environmentally rich curriculum you could imagine. It's all there. I feel most comfortable when I'm teaching in situ in the Garden. I have a tree. I can talk about it. I know about it. I have a broccoli plant, whatever it is. That's your lesson plan, as it were. I'm somewhat uncomfortable even with notes and other instructional aides in the classroom, but we do a blend of both.

It's an unintentional temporary community for six months. That is to say, this person didn't come to live with that person—they came to study agriculture and horticulture—but they are thrust together in this unintentional community, and they live on site and they share all domestic duties: shopping and cooking and cleaning.

There are marvelous opportunities for learning and growth in terms of the social dynamics of living together. People come in and train them in various aspects. Not just, this is how you cook, although we have that too, but some stuff about group dynamics: here are some typical cycles that you'll go through. The ethic at the apprentice program is, try to work through consensus. So there's trainings around that. But the eclectic nature of the apprentice group and the richness of the various life and professional experiences is a huge dynamic, value-added informal curriculum.

For a number of years I had a good almanac and I brought it in at the beginning of the program and said, "Okay," just on a volunteer basis on any given day, "someone find where you live. Show us where it is, tell us about it." Oral, cultural history. It gets people going. People feel free about sharing their background professionally and culturally. But beyond that, there's actually a structure that we've sort of helped them organize, where they have evening classes. Like if someone has had experience with agriculture in the tropics because they were in the Peace Corps, they have a venue to speak, slide shows. So it actually has some focus and direction.

It's a bear for them to self-manage and us to manage forty people. Everybody wants their individually tailored program and curriculum, which of course is impossible. But the richness, especially when it goes well, is just amazing. It's an education in itself. People always say, "How come you never go anywhere?" I say, "Well, I don't have to go anywhere. The whole world comes to me through the apprentices."

The apprenticeship itself is a very life-defining experience. You find it with some groups more than others, but one that comes to mind is the group of 1989. That group is in contact with each other. There are ongoing social friendships, very deep bonding, even though people are far flung around the world and in varying professions and lifestyles. We had our [apprenticeship program] fortieth-year celebration last year, and I joked we should have one session of people who got together romantically when they were at the Farm. And then we could have another one with those who stayed together and those who didn't. [laughs] There're a lot of relationships that spring up that are lifelong or not. But beyond that, professional business relationships are spawned at the Farm. That's really dominant over the years, lasting business and friendship relationships that were formed here.

Probably one of the leading [former apprentices] would be Cathrine Sneed. You've probably heard of her. She was essentially a counselor at the South San Francisco County Jail, back when she was in social work. She came and did the apprentice program and then turned it into a gardening program there, to use it as kind of a vehicle and metaphor for trying to teach the prisoners that if they could take care of these plants, they could take care of themselves. For a number of years they had a tree corps, arborist tree care of street trees, city trees, park trees, and they had contracts with the city of San Francisco. The idea was to take these folks when they got out of jail and place them in a job so there would be lower recidivism. And they have a remarkable success rate. And then she started what she called the Outside Project, which was a series of gardens on the outside. I would say, hands down, to me, she's the most spectacular apprentice. She's a dynamic woman. She was recently on the Oprah show and I guess Oprah gave her a ten-thousand-dollar tractor for their project, or something like that. She's a mover and shaker.

I have to say, though, that some of the people that I have the fondest memory of are not so much the eco-stars, of which there are quite a number, but just people who are doing something somewhere in a real basic, almost monastic, Zen-like way, just doing it without the glory and all that. They could be a backyard gardener, a schoolteacher, or a small farmer.

I have more unsung heroes. A guy named Joe Schirmer, who was an apprentice about ten years ago, could have been a professional surfer, but he didn't go that way. He did the apprentice program. He's now the proprietor of what's called Dirty Girl Farms. He was just this eighteen-year-old punk when he came, [laughs] and now he's one of the lionized organic producers, and rightfully so, in the Central Coast of California.

If you look at the academic clout of the institutions that apprentices have gone to with graduate and undergraduate degrees, it includes, but is not limited to, Harvard, Stanford, University of California, Los Angeles, Berkeley, Ohio State, Michigan, Princeton, Cornell—major institutions. And then they're at a place in their lives, often the mid-twenties to mid-thirties, where they have much more direction and focus than undergraduate students. There are no job placements here. They have to find jobs themselves, but they have that intelligence, drive, and capability to do that. The success of the program is largely about the caliber of the students that come. It's impressive. They're quick studies. They're very brave.

[Most program graduates find ways to make a living in some aspect of agriculture]. I would qualify that by saying that there's some kind of quotient between reasonable recompense and they're satisfied with the work. I mean, there's a joke. I know three agricultural jokes. And one of them is, how do you make a small fortune in agriculture? And the answer is, you start with a large one. I mean, people don't go into this because they want to make a lot of money. It is an issue, especially as people get older and want to have families and have health insurance.

What I enjoy most about the apprentices was seriously brought home [to me] to the point of tears coming down the aisles last summer at our fortieth-year celebration (even now I'm getting a little choked up), is to see what they go out and do. It's spine-tingling. It's demonstrable. It's palpable. In a small, on-the-ground way, this project has changed the face of horticulture and agriculture in this country, particularly organic agriculture. That's a good reason to come to work in the morning, to be able to contribute to that.

Patricia Allen

Photo: Jennifer McNulty

Patricia Allen is one of the nation's most prominent scholars on social aspects of food production, distribution, and access. She illuminates the changing food system as it affects and is shaped by conditions of labor, gender, and social inequality. Her publications include Together at the Table: Sustainability and Sustenance in the American Food System, *published in 2004 by Pennsylvania State University Press, an edited volume,* Food for the Future: Conditions and Contradictions of Sustainability, *published in 1993 by John Wiley Press, and numerous articles and book chapters. Allen also directs the world-renowned Center for Agroecology and Sustainable Food Systems (CASFS) at UC Santa Cruz. She oversees some twenty-five employees; a twenty-five-acre farm and two-acre garden; a seasonal campus produce stand and community supported agriculture program; a residential Apprenticeship in Ecological Horticulture that annually trains a diverse group of thirty-five to forty organic farmers, gardeners, and educators; a wide range of research projects; publications and events; educational offerings for undergraduate and graduate students as well as the general public, and ongoing fundraising efforts.*

My grandmother was a single mother whose husband had died when her youngest child was six years old. She ran her own farm, which was very unusual at the time. She had a truck farm— vegetable crops, apricots, and cherries. From a very early age, I remember seeing her work so, so very hard for so little, as huge farms were growing up around her. That inspired not only my interest in food and farming, but also social justice.

From the time I was eight years old, I worked on the farm during the summer to pay for school clothes. I mostly cut apricots for drying. It was monotonous, boring work and was highly gendered. It was women and girls who cut the apricots, and men and boys who picked the apricots and drove the tractor. So my interest in gender issues was also launched at a very early age.

Farm workers were hired to pick the fruit and do other farm work. One of them lived in the area, and he worked year-round. But then there were migrant workers as well. Labor was divided by ethnicity, not just by gender.

I went to a Catholic all-girls' high school. In the Catholic school, there was a lot of emphasis on thinking and on compassion and poverty issues, not in a really didactic, hit-you-over-the-head way, but you knew from a very early age that there was poverty and hunger in the world and that that wasn't necessary.

I don't know how I quite had the moxie to do this, in retrospect, but when I was a sophomore I observed that the all-boys' Catholic school had a really different math and science curriculum. I said that I wanted to take these classes, too, and because the girls' school didn't offer them, I went to the boys' school.

At the University of California, Berkeley, the education was absolutely phenomenal. I was able to take classes in political economy from the premier people in the field. I majored in political economy because I wanted to understand what the core issues were in shaping the socioeconomic world of food and agriculture. I remember the relief of my math and science classes because they involved formulas and "right" answers while the political economy courses did not.

I graduated from Berkeley with honors, but I didn't come from a professional or connected family, and was in a male-dominated field. I did not receive any mentoring or encouragement for getting a job in my field. I took an unrelated minimum-wage job and saved my money from working at that job so I could travel around the world.

[I remember] being warned about traveling in certain countries. I traveled third class from Nairobi to Wadi Halfa in Sudan, feeling like I had rarely been so accepted in my life as by the other people on the train. People shared whatever they had. When I continued to travel in Sudan and stayed in Khartoum for a bit, I learned quickly to never admire anything, because they would give it to you. No matter how little people had, they gave.

I remember one time—I was in the desert of Sudan, traveling by local bus, and the bus stopped at an outpost for a break. I sat down to have tea next to a man in full nomadic regalia complete with sword on his belt. We were not able to communicate through language or any shared cultural history. But there we were sitting down with others, drinking tea in the middle of the desert. I remember that moment as one of those most amazing human communication experiences that spans time, ethnicity, gender, politics. Just basic humanity. Experiences like that reminded me about the importance of generosity and of not stereotyping and categorizing people.

While I was traveling, I thought, well, there's so much going on in the world; what I really need to do is go to graduate school and learn how I can make more of a contribution. So I went to graduate school at UC Davis, and studied international agriculture. At that time I thought that I would go overseas and do international agricultural development work, inspired by my travels in Africa. As it happened, that's not how it went. Through my education in this graduate program, I realized that so much of what was going on was the United States developing ideas and programs and then exporting them and saying, "Be like us." It was also clear that "being like us" wasn't always appropriate. I decided that I would be better off working on domestic food and agriculture areas in which I had more direct experience.

I've never had the luxury of just going to school without also having to have a paid job. So I became a research assistant in the Small Farm Program at UC Davis. My role was to organize programs and collect information, to put out research briefs and things like that. Eventually, a regular staff position was created, for which I applied. I remember in the job interview, the question of why a woman would want to have a job outside the home came up. And I thought that question was really strange, but at the time I didn't have an analytical category or a context for understanding the ways in which it was inappropriate and irrelevant.

I remember organizing the first statewide Small Farm Conference, which has since become an annual conference. It felt very exciting to work with other people to do something that hadn't been done before, helping people access the knowledge located within the University of California. The conference was about direct marketing and crop production. It was designed to help small farmers increase their success at an individual farm level. It wasn't issue-oriented.

At the Small Farm Center, we had ideas about what would be useful information for small farmers, but there was no systematic study of their needs. So that's what I did for my master's thesis: a statewide survey of California small-scale growers to find out what they really needed from the University in terms of research and education.

Those were very different times for women in agricultural professions. I was frequently the only woman at meetings I attended for the Small Farm Center. I remember one time going through the cooperative extension directory at UC Davis and getting the sense that that all of the support staff were women, and all of the specialists and farm advisers, except for in home economics, were men. This gendered division of labor has changed significantly in the food and agricultural sciences although there's still a long way to go to achieve gender equity.

I really enjoyed the constellation of expertise that existed at UC Davis. I learned a lot. I also remember participating in the early days of the Student Farm. It felt like something was being created that was going to add to the work being done in the existing agricultural research and education departments. I'm happy to say that the UC Davis Student Farm has grown since then and is flourishing today.

I arrived at UC Santa Cruz in early 1984. At that time, sustainability was a controversial topic. Just raising that as a topic for the Small Farm Conference was met with skepticism, whereas at what was then called the Agroecology Program at UC Santa Cruz, it was the core of our work. It was exciting to be part of the early days of sustainability.

At that time, sustainability tended to be mostly focused on environment. People's basic needs and social justice were not a key part of the equation, at least the way sustainability was being conceptualized in the U.S. After being here for several years, I wanted to learn more about social theory so I did what I had always done, working and going to school at the same time. I applied and was admitted to the graduate program in the UC Santa Cruz Sociology Department.

It was an amazing education. I would go to professional conferences and feel so fortunate to be among the cutting-edge thinkers at UC Santa Cruz. Through the coursework and seminars in the sociology program I was able to deepen my understanding of sustainable agriculture and environmental and social issues. Through the sociology grad program I developed a deeper understanding of the theoretical and applied aspects of gender, class, and race issues. While I didn't take any classes on the food system—there weren't any at that time—I was able to apply what I was learning, in general terms, to what was going on in the food system. This has formed the foundation of my research and writing on sustainable food systems.

Eventually I became associate director [of the Center], with the responsibility for the social science research program. Then I became interim director, [and eventually] the dean asked if I would stay on as director. The job is very complex because we have a large staff, multiple and diverse programs, farm and garden facilities, a residential program, and most of our more than two-million dollar budget is extramurally generated. It's a complicated organization, with lots of history and lots to contribute to improving the food system.

Right now I feel that the scope of the food crisis is almost unprecedented. And so is the interest in sustainable food systems. We have this intersection of crisis and interest. There's never been a more important time to be working on sustainable food systems. There's such a need for expanding what it is that we do and can do, for making more of a difference in the world than we already are. Food is something with which everybody is engaged, every day. I mean, if you're lucky, you eat every day. Everybody thinks about food every day. It seems to me to be such a locus of social change and understanding—understanding how global warming is affected by the choices that people make in what they decide to eat, for example.

Already our apprenticeship program has had a huge influence. People have gone on to create their own programs, or increase the effectiveness of programs in California, and nationally and internationally. Some of the natural science research we've done has been absolutely path-breaking in terms of legitimizing doing regular scientific research on organic farms. Land-grant universities throughout the country are now creating student farms, organic research programs. Today there's something like 150 sustainable agriculture programs or sub-programs in American universities; probably ten years ago there were close to zero. Many cite the Center.

It's rare anymore that people don't include social justice in a definition of sustainability, whereas when I was starting out, it seemed that there was resistance [to that idea]. We've always been on the cutting edge [at CASFS]. It's a role that I feel is our role to play. My hope for the Center is that it continues as a place where we are engaged in innovative research and discourse, as well as teach people the practical skills that they need to flourish in the food system. Knowing how to grow your own food is so important!

When I was in academic courses or at scholarly conferences, I would think, we really need to get out there and do something, not just think! And then I'd be in these meetings of NGOs and I'd think, we really need theory to inform our actions! The intersection of the theoretical and the practical, all moving in the direction of environmental soundness and social justice—to me that is the goal.

And that we change the world. And that no child ever goes hungry, ever. That definitely drives me.

Research is important because it's the way in which we discover new ways of doing and thinking about things. Without research and innovation, all we could do is to keep walking along the same path that has gotten us into the situation that we're in right now. It's interdisciplinary research because the agri-food system, of course, is interdisciplinary. There's producing on farms; there's farm labor relationships; there's who owns the land; there's what are consumers thinking about, how do they make their choices, how do they decide how they're going to intervene in the food system; there's public policy, which determines, largely, what kind of a food system we have in this country; there's hunger in this country and in other countries; and then there's the aesthetics of food and the art of food. So it really does encompass all the disciplines.

At the Center's strawberry festival, UC Santa Cruz students organized a workshop on strawberries and social justice. A student said, "One of the issues is that farmers need to make more money, and so a lot of them are growing organic strawberries, but then, on the other hand, you hear that organic food is so expensive that low-income people can't afford it. So how does this get resolved?" This student is raising a key issue in developing sustainable food systems, and asking us to open up what we think is possible, to look at new ways of addressing the problems we face.

For example, rather than assuming that people's access to food has to happen through a market mechanism, which is determined by income, we could be looking at other ways of making sure that people have food. We could turn the discourse around food from one that is based on ability to pay, to one of human rights, a discourse of entitlement. That really changes the way you even think about the questions that need to be addressed.

Another example is that when we talk about sustainable food systems, a lot of the emphasis is on what farmers need and the production system. When you look at it, about 7 percent of the people who work in the food system in this country are working in production itself. Of that percentage, very few are farmers; the rest are hired farm workers. We rarely talk about food processing workers, food-service workers, or food transportation workers. So we can change how we shape and frame questions and increase efforts to reduce the environmental and human costs of the food system.

Farmers need to make higher incomes, which raises the price of food, which then makes it less accessible to low-income people. This may be case with alternative food distribution networks such as farmers' markets and community supported agriculture, even though the leaders of these organizations are committed to increasing access for low-income people. The questions are what range of options are available and who will pay the costs?

Aside from the economic barriers to participation in alternatives for fresh, local food, there are also cultural barriers. People don't necessarily feel comfortable—some people have said that getting a box from a CSA feels like what you got from the food bank. People that participate in community supported agriculture tend to be affluent, European American, highly educated people. People are working hard to build on this base to include greater diversity and expand access.

The question of whether labor justice should be a criterion for organic certification is an interesting one, and I can argue both angles. In my mind, what we're trying to do is make the food system more sustainable, and I include social justice as a key component of sustainability. So, on the one hand, it doesn't make much sense to me that we would have certification criteria for only one aspect of sustainability. On the other hand, it doesn't make sense to hold organic growers to a standard that could price them out of the market. The solution requires new ways of doing things and has to be contextualized in terms of everything we're trying to change about the food system.

Some of our preliminary research on preferences for different kinds of labels has found that people are interested in labels that would have to do with living wages for workers along with labels for local and organic food. Certification efforts are beginning to include social justice criteria as well as environmental criteria. Maybe they don't go far enough. Maybe not everyone will be able to afford it. But regardless, these efforts are inspiring people to think more deeply about their food and the food system as a whole. I think we need models on the ground that illustrate how things could be different. There will be opportunities to create policies that foster sustainable food systems.

There are many inspiring and innovative people and programs working for sustainable food systems. And because of a number of crises that have been in the press, and because of some of the popular publications and movies, there is a phenomenal public interest right now in the food system. People are really starting to pay attention. What information they get, I think, is going to determine the choices that they make. The environmental and equity issues have to come together..

I think the strength of food-system localization is that people can see right in front of them what the food system looks like, who's participating, who's winning, who's losing. That is a real strength for community engagement. Certainly if you are interested in eating in season, it will become very visible to you what you can eat at any given time of the year. At the same time, we need to think and act beyond our own communities to create sustainable food systems without borders.

Today so many people in so many places around the world are working on "getting it right" about the food system. People from so many different walks of life are interested and engaged. Food is the one thing we produce that everybody needs. We can't afford to not have a sustainable food system.

Selected Bibliography

Allen, Patricia. *Food for the Future: Conditions and Contradictions of Sustainability*. New York: Wiley, 1993.

————. *Together at the Table: Sustainability and Sustenance in the American Agrifood System*, Rural Studies Series. University Park, PA: Pennsylvania State University Press, 2004; published in cooperation with the Rural Sociological Society.

Allen, Will. *The War on Bugs*. White River Junction, VT: Chelsea Green Publishing, 2008.

Berry, Wendell, and Sierra Club. *The Unsettling of America: Culture & Agriculture*. 3rd ed. San Francisco: Sierra Club Books, 1996.

Chadwick, Alan, and Stephen J. Crimi. *Performance in the Garden: A Collection of Talks on Biodynamic French Intensive Horticulture*. Mars Hill, NC: Logosophia Press, 2007.

Conford, Philip. *The Origins of the Organic Movement*. Edinburgh: Floris Books, 2001.

Mary V. Gold and Jane Potter Gates. *Tracing the Evolution of Organic/Sustainable Agriculture: a Selected and Annotated Bibliography*. Beltsville, MD: United States Department of Agriculture, National Agricultural Library, [1988]; updated and expanded, May 2007, Internet-resource: http://www.nal.usda.gov/afsic/pubs/tracing/tracing.shtml

Gliessman, Stephen R. *Agroecology: The Ecology of Sustainable Food Systems*. 2nd ed. Boca Raton, FL: CRC Press, Taylor & Francis Group, 2007.

Guthman, Julie. *Agrarian Dreams: the Paradox of Organic Farming in California*, California Studies in Critical Human Geography, 11. Berkeley: University of California Press, 2004.

Henderson, Elizabeth, and Robyn Van En. *Sharing the Harvest: A Citizen's Guide to Community Supported Agriculture*. Rev. enl. ed. White River Junction, VT: Chelsea Green Publishing, 2009.

Howard, Albert. *An Agricultural Testament*. New York, London: Oxford University Press, 1943.

Imhoff, Dan, Jo Ann Baumgartner, and Wild Farm Alliance. *Farming and the Fate of Wild Nature: Essays in Conservation-Based Agriculture*. 1st ed. Healdsburg, CA.: Watershed Media, 2006.

Jolly, Desmond Ansel and Isabella Kenfield. *California's New Green Revolution: Pioneers in Sustainable Agriculture*. Regents of the University of California Small Farm Program, 2008.

Jolly, Desmond Ansel, editor. *Outstanding in Their Fields: California's Women Farmers.* Davis, CA: UC Small Farm Center, 2005.

Kimbrell, Andrew. *The Fatal Harvest Reader: The Tragedy of Industrial Agriculture.* Washington, D.C.: Island Press, 2002.

Kingsolver, Barbara, Steven L. Hopp, and Camille Kingsolver. *Animal, Vegetable, Miracle: A Year of Food Life.* 1st ed. New York: HarperCollins Publishers, 2007.

Kristiansen, Paul, Acram Taji, and John P. Reganold. *Organic Agriculture: A Global Perspective.* Ithaca, NY: Comstock Publishing Associates, 2006.

Lockeretz, William. *Organic Farming: An International History.* Wallingford, UK and Cambridge, MA: CABI, 2007.

McNamee, Thomas. *Alice Waters & Chez Panisse: The Romantic, Impractical, Often Eccentric, Ultimately Brilliant Making of a Food Revolution.* New York: Penguin Press, 2007.

Merrill, Richard. *Radical Agriculture.* New York: New York University Press, 1976.

Nabhan, Gary Paul. *Coming Home to Eat: The Pleasures and Politics of Local Foods.* 1st ed. New York: W.W. Norton and Company, 2002.

Northbourne, Walter James. *Look to the Land.* 2nd, rev. special ed. Hillsdale, NY: Sophia Perennis, 2003.

Pollan, Michael. *The Omnivore's Dilemma: The Search for a Perfect Meal in a Fast-Food World.* London: Bloomsbury, 2006.

Robinson, Jennifer Meta, and J. A. Hartenfeld. *The Farmers' Market Book: Growing Food, Cultivating Community.* Bloomington, IN: Quarry Books/Indiana University Press, 2007.

Shiva, Vandana. *Soil Not Oil: Environmental Justice in a Time of Climate Crisis.* Cambridge, MA: South End Press, 2008.

Smith, Alisa, and J. B. MacKinnon. *Plenty: One Man, One Woman, and a Raucous Year of Eating Locally.* 1st U.S. ed. New York: Harmony Books, 2007.

Somers, Robin, ed. *The Chadwick Garden Anthology of Poets,* Santa Cruz, CA: Somersault Studios (A Project of the Friends of the UCSC Farm and Garden and the Center for Agroecology and Sustainable Food Systems (CASFS)), 2009.

Van den Bosch, Robert. *The Pesticide Conspiracy.* Berkeley: University of California Press, 1989.

Walker, Richard. *The Conquest of Bread: 150 Years of Agribusiness in California.* New Press, 2004; Distributed by W.W. Norton and Company.

Timeline

This timeline traces key moments in the development of organic farming and sustainable agriculture, on the Central Coast of California, in the United States, and beyond.

1911: American agronomist F. H. King publishes *Farmers of Forty Centuries, Permanent Agriculture in China, Korea, and Japan*, still an important reference on the farming practices of ancient cultures.

1924: A group of European farmers, concerned with the decline of soil quality and crop health as a result of the use of chemical fertilizers, seek advice from Dr. Rudolf Steiner, founder of anthroposophy (a philosophy based on the view that the human intellect has the ability to contact spiritual worlds). Steiner gives lectures on biodynamic agriculture at Koberwitz, Germany, in June 1924. From these talks, Steiner's fundamental principles of biodynamic farming and gardening emerge, which Alan Chadwick later fuses with French intensive gardening methods at UC Santa Cruz.

1938: Microbiologist Masanobu Fukuoka resigns his job as a research scientist to devote his life to the development of the "no-till" organic method of growing grain. In 1975, Fukuoka publishes *One Straw Revolution*.

1938: The United States Department of Agriculture Yearbook of Agriculture publishes *Soils and Men*, a manual on organic farming still used today. That same year the pesticide DDT is invented by a Swiss chemist. It is first used on a farm in the United States in 1942.

1939: Lady Eve Balfour of Suffolk, England, one of the first women to study agriculture at an English university, conducts her groundbreaking Haughley Experiment in which she compares organic farming and conventional chemical farming on two adjoining farms. In 1942, Balfour publishes the initial findings in her book, *The Living Soil*. In 1946, she cofounds the Soil Association, an organic advocacy group, which is the major organic farming organization in the United Kingdom today.

1940: British agronomist Lord Walter James Northbourne publishes *Look to the Land*, which describes farms as organisms, advocates an ecologically-balanced approach to farming, and uses the word "organic" to describe a sustainable agricultural system.

1941: Twenty million Americans plant Victory Gardens during World War II.

1942: The United States and Mexican governments start the Bracero Program, which brings 4.5 million Mexican nationals to work in the fields in the United States. This sometimes exploitative and controversial program addresses labor shortages in the fields and continues until 1964.

1942: Jerome Rodale begins publishing *Organic Farming and Gardening* magazine.

1943: Sir Albert Howard, British mycologist and agricultural researcher, publishes *An Agricultural Testament*, based on research in India. This book becomes a classic text on soil fertility. Howard is considered a founder of the organic farming movement.

1945: Nerve gas research during World War II results in a new class of pesticides which begin to be widely used in the post-war period.

1948: American writer Louis Bromfield publishes *Malabar Farm*, which describes his experiment in sustainable farming in Mansfield, Ohio.

1962: Biologist Rachel Carson publishes *Silent Spring*, a landmark environmental book which documents the negative impact of agricultural chemicals.

1965: Mothers in Japan, concerned about the rise of imported food and the loss of arable land, start the first community supported agriculture (CSA) project.

1965: Fred Rohé opens New Age Natural Foods in San Francisco, one of the first natural foods stores in the United States.

1965: El Teatro Campesino ("farmworkers' theater") is founded as a cultural arm of the United Farmworkers. The original actors are all farmworkers and productions are held on flatbed trucks in the middle of the fields.

1966: Cesar Chavez, Dolores Huerta, and others form the United Farm Workers Organizing Committee, which in 1972 becomes the United Farm Workers Union (UFW).

1966: The San Francisco Diggers begin giving away free food.

1967: English master gardener Alan Chadwick is hired to create a Student Garden Project on the UC Santa Cruz campus.

1967-1970: Grape strikers from the UFW organize a successful international grape boycott.

Late 1960s: People organize to purchase produce, dry goods, eggs, and other food directly from farmers and small distributors, in what becomes known as food conspiracies. These are usually affiliated with the New Left movements of the time. Some food conspiracies evolve into food co-ops and natural food stores.

1970: (April 22) Earth Day. Founded by Senator Gaylor Nelson (D-Wisconsin). Twenty million Americans demonstrate for a healthy, sustainable environment.

1971: Chez Panisse Restaurant is opened by Alice Waters in Berkeley, California. Chez Panisse serves local, organic foods, creating a market for organic produce in the Bay Area. Chez Panisse is credited with inventing California Cuisine.

1971: Rodale Press's *Organic Farming and Gardening Magazine* initiates an organic certification program in California. This program ultimately inspires the formation of California Certified Organic Farmers.

1971: The Maine Organic Farmers and Gardeners Association (MOFGA) is formed and is the oldest and largest state organic organization in the country.

1971: Northeast Organic Farming Association of Vermont (NOFA) establishes organic certification standards.

1972: The International Federation of Organic Agriculture Movements (IFOAM) begins in Versailles, France during an international congress on organic agriculture organized by the French farmer organization Nature et Progrès.

1973: Dan Janzen publishes "Tropical Agroecosystems" in *Science* magazine, a seminal article in the development of agroecology.

1973: California Certified Organic Farmers (CCOF) is founded.

1974: Warren Weber founds Star Route Farms in Bolinas, California, now the oldest continuously certified organic grower in California.

1974: Veritable Vegetable, now the nation's oldest distributor of certified organic produce, opens in San Francisco, California.

1974: Oregon Tilth certification agency is founded.

1975: CCOF's first chapter, the Central Coast Chapter, is formed by Santa Cruz County members.

1976: Wes Jackson founds The Land Institute in Salina, Kansas, an organization which, according to its website, continues to devote resources "to developing an agricultural system with the ecological stability of the prairie and a grain yield comparable to that from annual crops."

1976: Ecologist and horticulturalist Richard Merrill edits *Radical Agriculture*, a formative text in the sustainable agriculture movement.

1977: UC Santa Cruz Farm apprentice Stephen Decater, and his wife, Gloria, start Live Power Community Farm in Covelo. Decater studied with Alan Chadwick at the Round Valley Garden Project from 1972 to 1974.

1977: Wendell Berry publishes *The Unsettling of America: Culture and Agriculture*, a devastating critique of industrial agriculture.

1978: Governor Jerry Brown signs the Direct Marketing Act, allowing California farmers to sell their produce directly to consumers at locations designated by the Department of Agriculture. This is a boon to farmers' markets in California.

1978: Life Lab project starts at Green Acres School in Live Oak (near Santa Cruz) California. Life Lab teachers transform a vacant lot into a thriving garden and develop curriculum based on the idea that children are motivated to learn scientific ideas by asking questions while in the garden.

1978: The California Agrarian Action Project (CAAP) forms in Yolo County, California to organize demonstrations in support of farm workers in dire economic straits due to unemployment partially caused by adoption of the mechanical tomato harvester.

1979: CAAP files a landmark suit against the University of California for using taxpayer dollars to develop technologies that benefit large farms and hurt small farms and farm workers. This becomes known as the Research Priorities (or "Tomato Harvester" or "Mechanization") case.

1979: The California Organic Food Act is signed into law. While it is a state-mandated local program, no budgetary appropriations are allocated for enforcement. Any infractions have to be taken up in the courts by organizations like CCOF.

1980: USDA publishes the *Report and Recommendations on Organic Farming* to increase "communication between organic farmers and the U.S. Department of Agriculture." The report is soon suppressed by the incoming Reagan administration.

1980: The original Whole Foods Market opens in Austin, Texas with a staff of nineteen people.

1981: Stephen R. Gliessman is hired by the UC Santa Cruz Environmental Studies Board as a faculty member. Gliessman founds the Agroecology Program.

1981: The first Ecological Farming Conference (organized by Amigo Bob Cantisano) is held at the Firehouse in Winters, California. Forty-five people attend. The conference is now the largest sustainable agriculture gathering in the Western United States.

1981: The Ecological Farming Association is founded. In addition to organizing the annual EcoFarm Conference, EFA puts on training programs, on-farm events, and communications initiatives to "pursue a safe and healthful food system that strengthens soils, protects air and water, encourages diverse ecosystems and economies, and honors rural life." EFA is located in Watsonville, California.

1982: The Alfred E. Heller Chair in Agroecology is founded with a $375,000 gift from Alfred E. Heller and is the first endowed chair at UC Santa Cruz.

1983: California Association of Family Farmers (CAFF) is founded.

1983: Sibella Kraus and others organize the first Tasting of Summer Produce festival in the San Francisco Bay Area.

1984: Jan Vander Tuin brings the concept of community supported agriculture (CSA) to North America from Europe. Vander Tuin had cofounded a community supported agricultural project named Topinambur, located near Zurich, Switzerland.

1985: The Rural Development Center (RDC) is founded on a farm eight miles south of Salinas, California in 1985 by the Association for Community-Based Education (ACBE) of Washington, D.C. The RDC pioneers the idea of a "Farmworker to Farmer" program where agricultural workers gain broader skills leading to their advancement. This program eventually grows into the Agriculture & Land-Based Training Association (ALBA).

1986: After a statewide organizing effort from CAFF and other activists, Senate Bill 872, which creates the University of California's Sustainable Agriculture Research and Education Program (UC-SAREP), passes in the California legislature.

1986: Slow Food begins in Italy when a group of organizers form Arcigola to resist the opening of a McDonald's in Rome. Carlo Petrini transforms Arcigola into Slow Food, an organization devoted to preserving the cuisine and seeds indigenous to an eco-region. The movement now extends globally to over 100,000 members in 132 countries.

1987: CCOF publishes the first edition of the *CCOF Certification Handbook and Materials List* and the first *Farm Inspection Manual*, as well as organizing the first series of farm inspector trainings.

1987: Swanton Berry Farm becomes the first certified organic strawberry farm in California.

1988: Steve Decatur starts the first CSA program in California, at his Live Power Farm in Covelo.

1988: United States Department of Agriculture establishes the Sustainable Agriculture Research and Education (SARE) program.

1988: In cooperation with the California Department of Health Services, CCOF pursues an investigation of Pacific Organics, a distributor selling conventionally grown carrots as organic. This event becomes known as "The Carrot Caper."

1989: CBS's *60 Minutes* airs "Intolerable Risk: Pesticides in our Children's Food." This prompts what becomes known as "the Alar scare." Meryl Streep, Hollywood spokesperson for the Natural Resources Defense Council, appears on The Phil Donahue Show supporting local farms and organic foods.

1989: Amigo Bob Cantisano opens Organic Ag Advisors. Cantisano is the only professional organic ag adviser in the state of California at that time.

1990: California Governor George Deukmejian signs the California Organic Food Act (COFA), authored by California State Assemblyman Sam Farr. COFA establishes standards for organic food production and sales in California and becomes one of the models for the National Organic Program's federal organic standards.

1990: The Federal Organic Foods Production Act (OFPA) is completed as part of the Farm Bill. This act sets out to establish national standards governing the marketing of organically produced products, assure consumers that organically produced products meet a consistent standard, and facilitate interstate commerce in both fresh and processed organic foods.

1990: Citizens Committee for the Homeless, a Santa Cruz County nonprofit, founds the Homeless Garden Project.

1990: UC Santa Cruz Agroecology Program researcher Sean Swezey, Steve Gliessman, and others publish the first organic strawberry conversion study. The study is conducted on Jim Cochran's Swanton Berry Farm.

1992: The Organic Farming Research Foundation (OFRF) is formed as a spinoff of CCOF, to fund the educational objectives of CCOF and on-farm research of organic growing practices.

1992: The Organic Farming Research Foundation conducts the first of four National Organic Farmers' Surveys.

1993: UC Santa Cruz Agroecology Program's name is changed to the Center for Agroecology & Sustainable Food Systems.

1993: The first of five biannual Organic Leadership Conferences is held at the Claremont Hotel in Berkeley. These conferences are formative to the organic movement and are organized by the Organic Farming Research Foundation.

1995: First Western Region Community Supported Agriculture conference takes place in San Francisco, California.

1997: Steve Gliessman publishes *Agroecology: Ecological Processes in Sustainable Agriculture*, the first textbook of its kind.

1997: Mark Lipson and the Organic Farming Research Foundation publish *Searching for the "O-Word": An Analysis of the USDA Current Research Information System (CRIS) for Pertinence to Organic Farming.* The report documents the absence of publicly funded organic research at a critical political moment in the trajectory of the organic farming movement.

1998: Swanton Berry Farm becomes the first organic farm to sign a contract with the United Farm Workers.

1998: California FarmLink is founded to "build family farming and conserve farmland in California by linking aspiring and retiring farmers; and promoting techniques and disseminating information that facilitate intergenerational farm transitions."

1999: The International Agroecology Short Course is founded by Steve Gliessman.

2000: Wild Farm Alliance, a nonprofit organization dedicated to increasing biodiversity by expanding the idea and practice of wild farms, is founded in Watsonville, California.

2000: UC Santa Cruz Center for Agroecology and Sustainable Food Systems researcher Sean Swezey publishes the first organic apple production manual, based on research done in collaboration with apple farmer Jim Rider.

2000: After ten years of political debate, the final National Organic Standards rule is published in the *Federal Register* on December 21, 2000, establishing United States Department of Agriculture standards for organic food in the United States.

2001: Community Agroecology Network (CAN) is founded by Steve Gliessman and Robbie Jaffe.

2001: The Organic Farming Research Foundation publishes Jane Sooby's *State of the States: Organic Systems Research and Land Grant Institutions.* The report catalogues organic research, education, and extension projects in place at the nation's public land grant agriculture schools and at public research stations and through Cooperative Extension.

2002: CAFF's Hedgerow and Habitat project begins on the Central Coast, the North Coast, and in Stanislaus County. The program demonstrates the importance of growing native plant hedgerows to provide habitat for wildlife and beneficial insects.

2002: Governor Gray Davis signs the California Organic Products Act (COPA). Beginning January 1, 2003, all products sold in California with less than 70 percent organic ingredients are not allowed to use the word "organic" on the front panel. However, in 2003, the State Assembly repealed the non-food provision of COPA.

2002: Swanton Berry Farm is awarded the Environmental Protection Agency's Stratospheric Ozone Protection for developing methods of growing strawberries without the use of methyl bromide, a chemical which depletes the earth's ozone layer.

2004: University of California students from the California Student Sustainability Coalition meet at UC Santa Barbara to launch the UC Sustainable Foods Campaign.

2006: An outbreak of food-borne illness caused by the pathogen *E. coli* 0157:H7 sickens about two hundred people and kills three; the outbreak is traced to bagged fresh spinach grown in San Benito County and the source is ultimately found to be cow manure.

2006: Michael Pollan publishes *An Omnivore's Dilemma: A Natural History of Four Meals.*

2007: Youth members of The Food Project, the California Student Sustainability Coalition, and other organizations meet at the Food and Society Conference sponsored by the W.K. Kellogg Foundation to found The Real Food Challenge, a national organization working to unite students for just and sustainable food.

2007: Barbara Kingsolver publishes *Animal, Vegetable, Miracle: A Year of Food Life,* which helps galvanize the locavore movement, as does Alisa Smith and J.B. MacKinnon's *Plenty: One Man, One Woman, and a Raucous Year of Eating Locally.*

2008: Larry Jacobs of Jacobs Farm/Del Cabo wins a landmark pesticide drift case against pesticide application company Western Farm Service, Inc. The court finds that the contamination of organic crops caused by pesticides drifting after application violated the rights of the organic crop grower. In 2010, Jacobs wins this case on appeal.

2009: President Barack Obama appoints Kathleen Merrigan, an academic and former congressional aide who helped write federal organic food-labeling rules, to be Deputy Secretary of Agriculture.

2009: First Lady Michelle Obama creates an organic vegetable garden at the White House.

2010: Mark Lipson accepts post as Program Specialist for Organic Farming for the United States Department of Agriculture.

2011: More than one thousand EcoFarm attendees sign a petition protesting the United States Department of Agriculture's decision to allow the planting of genetically modified alfalfa.

Index

Note: page numbers **in bold** refer to that person's oral history.

UCSC Farm Steps. Photo: Laura McClanathan

About the Project Team

Ellen Farmer, Irene Reti, and Sarah Rabkin (l-r) at the celebration
for Cultivating a Movement, UC Santa Cruz Farm, 2010

Project Director, Interviewer, and Editor Irene Reti directs the
Regional History Project at the UC Santa Cruz library, where she has
worked as an editor and oral historian since 1989. She holds a B.A. in
Environmental Studies and a Master's in History from UC Santa Cruz.
Her novel, *Kabbalah of Stone,* was published in 2010.

Interviewer and Editor Sarah Rabkin has taught in UC Santa
Cruz's writing program and environmental studies department for over
twenty-five years. She has led an undergraduate seminar for the Program
in Community and Agroecology that focuses on concepts of commu-
nity and agroecology in the context of sustainability. She holds a B.A. in
Biology from Harvard University and a graduate certificate in Science
Communication from UCSC. Her book of essays, *What I Learned at Bug
Camp*, was published in 2011.

Interviewer Ellen Farmer has a B.A. in journalism from San Jose State
University and a Master's in public policy from the Panetta Institute at
California State University, Monterey Bay with a specialization in issues
in sustainable agriculture, particularly coffee growing. Farmer worked on
an interim basis as marketing director for the California Certified Organic
Farmers in 2006.